The AgeTech
Revolution

The AgeTech Revolution

A Book about the Intersection
of Technology and Aging

KEREN ETKIN

M.A Gerontology

Founder of TheGerontechnologist

NEW DEGREE PRESS

COPYRIGHT © 2021 KEREN ETKIN

The AgeTech Revolution

A Book about the Intersection of Technology and Aging

ISBN   978-1-63730-706-9   *Paperback*

         978-1-63730-797-7   *Kindle Ebook*

         979-8-88504-009-9   *Ebook*

To my grandparents

Contents

"Get old. You're lucky if you do."

JANE CARO, "GROWING OLD: THE UNBEARABLE
LIGHTNESS OF AGEING" TEDXSOUTHBANK, 2015

Introduction

———

When Jake's grandfather passed away, his eighty-seven-year-old grandmother found herself living alone in a big home she didn't really need and couldn't afford to maintain. For many older people looking to downsize and move to a smaller home while having the option to make new social connections, a retirement community is the go-to solution. Jake's grandmother, however, decided to move into a smaller apartment in the same neighborhood that she knew and loved, and Jake was there to assist with everything she needed—from moving arrangements to helping her with rides to the doctor and grocery shopping. He soon realized there must be many other older adults just like his grandmother—people who are physically and cognitively fully able to take care of themselves, who wish to downsize, but don't want to be alone. Thus, he created UpsideHom, a fully managed apartment service for people fifty-five and older who are looking for an alternative to senior living with companionship built in.

The concept of UpsideHom is powerful because it taps into an important trend in aging—aging in place, which means that most older adults prefer to continue living in the community they know and love, rather than move to a retirement community. The reality is we probably need millions of Upside-Hom units to support the millions of adults who will reach old age in unprecedented numbers over the next few decades. As a gerontologist working in start-ups and writing about tech for older adults, I wondered how technology was poised to solve some of the biggest challenges and opportunities in aging, and what I have found gives me hope that technology offers a powerful future as we age.

I started my career in the "aging industry" working in the nonprofit sector. I worked with older adults from different socioeconomic backgrounds to tackle the different challenges they were facing in their everyday lives, from the most basic needs of food and medications to social engagement and recreational activities. There was really no time to discuss technology because there was always something more urgent to tend to.

That all changed for me when I was recruited as the first employee to an early-stage ambitious startup called Intuition Robotics. We were on a mission to tackle one of the biggest challenges of aging—social isolation and loneliness—with a social robot. At this stage, some of you might be thinking: "Why in the hell would you want to replace humans with robots?" This wasn't the purpose, and we'll get to that in Chapter Four.

Suddenly, talking to older adults about technology and reading every academic paper I could get my hands on about tech adoption by older adults and social robots for their demographic, became my full-time job. What I learned changed my worldview and my career trajectory forever.

Some people assume older adults don't want to use technology, but this couldn't be further from the truth. They do; however, they have different wants and needs than younger people. Some of it is a generational thing, some of it is a life-stage thing, and some of it depends on physical and cognitive changes that naturally occur as people grow older. The thing is most of the current technology available is not designed with older adults in mind as the potential end users. This is despite the fact they hold a significant portion of the world's wealth. The global spending power of people aged sixty and over was expected to reach $15 trillion by 2020 (Nahal and Ma, 2014). Older adults also spend quite a lot on tech products, and the global COVID-19 pandemic has caused many to ramp up tech spending.

The American Association of Retired Persons (AARP) report from April 2021 revealed tech spending among older adults almost tripled between 2019 and 2020, from $394 to $1,144 (Kakulla, 2021). However, most of the user-facing technology today is designed by younger adults and for younger adults, and older adults are almost never included in the design process or even considered as potential users.

Many people, especially digital natives who perceive themselves as very tech savvy, ask me: Why do we need to work on special solutions for older adults? After all, the people who don't use technology today will be replaced within a few decades by a new generation of older adults in which everyone will know how to use technology. In other words, they are saying, I'm sure that when I'm older, I'm going to be able to master technology the same way I do today.

This assumption is false for two main reasons. One is there is a natural, age-related decline in physical and cognitive abilities that are required to use the technology that's available today—vision, hearing, dexterity, and memory all change, and not for the better, as you grow older. Second, let's assume in a few decades, we will have "solved" biological aging—and humans will be able to choose whether or not they wish to go through the physical process of aging. Let's also assume most people won't need to proactively interact with technology in the future. User interfaces won't exist because technology will be ambient—it will be able to sense what we need and then execute tasks accordingly without our having to take any conscious action. Even if all this happens, there will still be a generational and life-stage difference in what people want.

I am now in my thirties. I don't necessarily want to use products that teenagers are using. I'm in a different stage in my life and I have different wants and needs, motivations, and priorities. A similar change will happen when I'm in my eighties. I'm probably going to have different needs, motivations, and priorities from my thirty-year-old self and from people who are in their thirties, or forties, or even fifties. My point is **tech for older adults is important not just because of usability but also because of desirability**. It's about developing solutions that cater to people's wants and needs, not just their abilities or disabilities. It's also about viewing older people through a wide lens, as complete human beings, and not from the limited perspective of health care consumers or care recipients.

I made it my mission in life to change the fact many of the user-facing tech-enabled products and services being developed today do not take into account older adults in the design process. Over the past few years, I've worked in start-ups that developed cutting edge technology for older adults and elder care providers. I've also created a media platform that covers the global agetech ecosystem, taught courses, given keynotes, and participated in countless panels about tech for older adults. My whole professional world revolves around the ecosystem of tech solutions for older adults, which I like to call *agetech*.

To me, this ecosystem is the silver lining. In the past few years, we've seen more and more start-ups come up with different digital products and services designed to tackle the different challenges of aging. We've even seen some of the world's biggest tech companies develop new products, or features in existing products, to serve older adults. These companies have been working on just about everything you can imagine, from robotics applications like autonomous cars and exoskeletons to computer vision applications like smart goggles that can read text aloud and identify objects in front of you. It's also worth mentioning that voice-enabled technology, which has become ubiquitous in recent years, is considered by many, including myself, to be a good example for an age-inclusive user interface we already have at our disposal today.

I'm optimistic, not just because it's my nature, but because I have seen firsthand some of the best and brightest minds in the tech world developing the most cutting edge solutions for the population that has been most underserved by technology: older adults. I believe we're on the verge of a revolution—the *Agetech Revolution.*

WHY SHOULD YOU CARE?

Our world is going through an unprecedented demographic shift.

By 2050, we'll have two billion people over the age of sixty living on this planet—that's twice as what we had in 2017 (UN, 2017). This is the result of two trends: a significant increase in life expectancy, and a significant decrease in birth rates. It means we'll not only have a larger-than-ever number of older adults alive, but they will represent an increasingly growing portion of the population. In the Organisation for Economic Co-operation and Development (OECD) member countries, people over the age of sixty-five are expected to be about 28 percent of the population by 2050 (OECD, 2017). We don't have to use our imagination to see what that's like; all we have to do is look to Japan, where one out of three people is already over the age of sixty (World Economic Forum, 2019).

This tectonic shift has implications on the macro level, of course, because it changes the dependency ratio (the number of people who are part of the workforce vs. those outside the workforce), which is an important metric for policymakers, insurers, employers, etc. However, on the micro level, if you're reading these lines and you have close family members over the age of eighty, chances are you will end up assuming the role of a family caregiver at some point in your life, if you haven't already. The "caregiver support ratio," which is the number of potential caregivers aged forty-five to sixty-four for each person aged eighty and older—is declining. Not too long ago, in 2010, every American over the age of eighty who needed care had seven potential family caregivers. By 2030, they will only have three (AARP Public Policy Institute, 2013). This has very real consequences for families, not only in America, but all over the world.

And while we haven't yet "solved" the care gap with technology, I believe in the next few decades we'll have at our disposal great solutions we can't even imagine today. After all, who would have imagined a watch that senses if you've fallen down and contacts emergency services, all by itself, twenty years ago?

Twenty years ago, all we had for helping older adults who were at risk of falling at home were personal emergency response systems (PERS), with clunky bracelets or pendants that had a "call for help" button. These days, we have a variety of home sensors and wearables to choose from that are able to not only detect if a person has fallen down but dispatch emergency assistance as well, without that person needing to press a button. My point is in 2021, we're in a much better position than we were ten or twenty years ago.

And while we haven't solved each and every challenge older adults are facing (not even close), we have many more commercially available solutions than ever before.

It's also important to note that, unlike some of the solutions that were common in the eighties and nineties, which, according to Coughlin and Yoquinto, are "big beige and boring" and made those who would wear them to feel old and frail, today we have fall detection solutions like the Apple Watch, which are designed to be beautiful and streamlined (Coughlin and Yoquinto, 2018). They make those wearing them feel cool and hip. This isn't to say that Apple is perfect when it comes to designing for older adults, but one has to admit that it's never embarrassing to be caught wearing or using one of their products—quite the opposite.

Why do Apple, Amazon, and other big tech companies, develop products and features for the older population? The answer is they recognize the demographics and are aware of the fact that older adults hold a significant amount of wealth.

The $15 trillion longevity economy has been noted (Nahal and Ma 2014). This number, $15 trillion, puts the longevity economy in the same ballpark as the United States and China, in terms of GDP. This is by far more wealth than millennials have ever held. Despite this fact, most of the products we see around us and most of the advertising is targeting millennials, who have less discretionary income than their parents or grandparents.

Policymakers from all over the world, from Canada, to the United States, to Australia, have realized the only way we can possibly solve the biggest demographic challenge the world has ever seen is with technology. Technology has played an increasingly bigger role in our lives in the past few years. Some even call this the fourth industrial revolution (Schwab, 2016).

We're surrounded by screens and notification sounds from the minute we wake up in the morning until the time we go to bed at night. In the past two years, during the COVID-19 outbreak and the vaccine rollout that followed, it became painfully clear that we can no longer exclude older adults from this revolution, and if we're not making a purposeful effort to include them in the digital transformation our society is going through, we're leaving people behind.

We have a huge opportunity to reshape our worldviews on tech and aging, on who technology is supposed to serve. The next five to ten years are critical for us as a society, and we need to make sure we are positioned to deal with the demographic shift ahead. It's also an opportunity to rethink who we're building tech for and what it provides: access, the ability to do everyday things like grocery shopping, and the ability to stay connected and engaged in society. This is especially true for entrepreneurs and investors, whether they are already part of the agetech ecosystem, or on the sidelines pondering whether they want to step in.

This ecosystem of companies developing technology for older adults, with its many wonderful startups, is still very much a blue ocean: "A market with little competition and barriers to entry" (Young, 2021). Those who capitalize on this opportunity and develop scalable solutions for the challenges of aging could potentially not only move the needle on these issues but also turn a significant profit.

In this book, I will explain why developing technology to tackle the challenges of aging is the single most important opportunity of the next decade. We'll discuss the underlying trends that impact this opportunity. Baby Boomers are retiring: this massive generation will all be over the age of sixty-five by 2030. The number of young caregivers is on the rise: Millennials and Gen Z make up 60 percent of new caregivers (Gordon, 2021). And of course, there are the ramifications of the COVID-19 pandemic. We'll explore key concepts like aging in place and the digital divide. We'll dive deep into some of the challenges of aging and how they impact the day-to-day lives of real people, with some notable examples. We'll also debunk three major myths and misconceptions about older adults and technology:

- First, that older adults have no interest in technology.
- Second, that older adults aren't using technology.
- Third, that you can't make money building products and services for older adults.

Lastly, I will try to make some predictions for the future. Where do I see the space in ten, twenty, or thirty years? As a gerontologist, an agetech entrepreneur, and someone who's been writing about the agetech ecosystem, I have come to a couple of realizations that I want to share with you. Are you ready?

CHAPTER 1:

What Is Aging?

———

Our world is going through an unprecedented demographic shift. Improvements in nutrition, sanitation, and medical care have enabled a significant increase in life expectancy all over the world (Costa, 2005). Never in human history have we had so many people reach old age. In just fifteen years, between 2015 and 2030, the number of people over the age of sixty in the world is projected to grow by 56 percent, from 901 million to more than 1.4 billion (UN, 2015). Currently, roughly one billion people over the age of sixty live on this planet, and many of them are healthy, active, and have discretionary income. The population over the age of sixty is growing faster than any other age group, and according to population projections, by 2050, 20 percent of the world's population (approximately two billion people) will be over the age of sixty (UN, 2017).

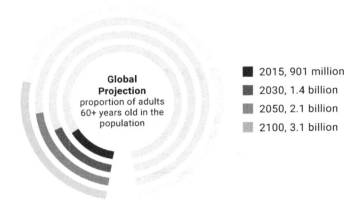

Global Projection
proportion of adults 60+ years old in the population

- 2015, 901 million
- 2030, 1.4 billion
- 2050, 2.1 billion
- 2100, 3.1 billion

Image source: TheGerontechnologist.com. Data sources:

1. United Nations, Department of Economic and Social Affairs, Population Division (2015). World Population Ageing 2015 - Highlights (ST/ESA/SER.A/368). https://www.un.org/en/development/desa/population/publications/pdf/ageing/WPA2015_Highlights.pdf

2. United Nations, Department of Economic and Social Affairs, Population Division (2017). World Population Ageing 2017 - Highlights (ST/ESA/SER.A/397). https://www.un.org/en/development/desa/population/publications/pdf/ageing/WPA2017_Highlights.pdf

3. United Nations, "World population projected to reach 9.8 billion in 2050, and 11.2 billion in 2100", 2017, https://www.un.org/development/desa/en/news/population/world-population-prospects-2017.html accessed September 2021

And it's not just the number of older adults that's expected to change drastically, it's also the economic implications. The longevity economy, which is the sum of all economic activity generated by people over the age of sixty, was expected to reach $15 trillion by 2020 (Nahal and Ma, 2014). In terms of GDP, this puts the longevity economy on the same scale as the United States and China, which currently have the two largest economies on earth.

WHEN DOES OLD AGE START?

When researchers ask people this question, they get different answers from people from different age groups, and not surprisingly, the older people get, the further away they push "old age" from their own age (Emling, 2017). As an individual's own age increases, their parameter for the beginning of old age is placed further along the lifespan (Kerrie et al., 2001). In a *Pew Research Center* survey, most respondents under the age of thirty said old age starts before sixty, whereas respondents over the age of sixty-five said old age starts at seventy-four (Taylor et al., 2009).

The OECD defines people over the age of sixty-five as "elderly". The United Nations High Commissioner for Refugees (UNHCR) states that the United Nations defines anyone over the age of sixty as an older person, and this age is also mentioned in the United Nations' declaration of the years 2021–30 as the *Decade of Healthy Ageing* (UN, 2021). However, in the UN's recent "World Population Ageing" report from 2020, it uses the age sixty-five to discuss older people.

Kowal and Peachey wrote in their 2001 World Health Organization (WHO) report, "In the developed world, chronological time plays a paramount role. The age of sixty or sixty-five years, roughly equivalent to retirement ages in most developed countries, is said to be the beginning of old age." Using retirement age as cutoff makes sense, because it acknowledges the major life change that usually occurs around it, but it is still somewhat arbitrary, since not everyone retires at sixty-five. It is also important to note not everyone that has crossed this age threshold holds the same characteristics, wants, and needs. In fact, the population over the age of sixty-five is quite diverse.

Anyway, since the average age of retirement in OECD countries is closer to sixty-five, this is the number I use when I refer to "older adults."

There's also no consensus in terminology—what is the correct term to call someone over the age of sixty or sixty-five? Seniors? Senior citizens? Elderly? Pensioners? Older adults? The English language has quite a few words to describe older people. In a *National Public Radio* (NPR) survey from 2014, respondents favored the term "older adults" (Jaffe, 2014). This term is also used in academic writing, and it is my personal favorite because I believe it is the most objective. Therefore, this is the term I use in my writing and will be using throughout this book.

WHAT IS AGING?

When you look up the word "aging" in the dictionary, you will find Oxford dictionary defines it as "the process of growing old."

This makes you realize we're all aging. Right now, as you are reading these lines, you too are growing old.

Merriam-Webster defines "old" in contrast between old and new: "Distinguished from an object of the same kind by being of an earlier date," Oxford does a similar thing with "Having lived for a long time; no longer young." (Lexico, 2021)

What does this process of growing old look like in 2021, and what is it going to look like in 2030 or 2050? Well, it's different for different people and different cultures, but the biological foundations are the same. According to Britannica's online encyclopedia, aging happens at the cellular, organ, and organism level.

Some of the typical signs of aging for humans, like graying hair and wrinkles, are external and visible. Others, like changes in your senses or cognition—vision, hearing, memory loss, processing speed, etc.—are internal (Harada et al, 2014). On top of physiological changes in aging, social and emotional changes are also associated with getting old.

Psychologist Tamara McClintock Greenberg wrote in an article for *Psychology Today* in 2013 that "growing older scares us for a variety of reasons. Some of us can't imagine being dependent. Others just want to be able to exercise like we did when we were twenty. For me, pain and sensory impairment just seemed like it would take away everything that matters to me—connecting with people I love as well as my value on being independent."

To help younger people understand what it means to live in an aging body, researchers at the Massachusetts Institute of Technology AgeLab developed AGNES: the Age Gain Now

Empathy System. It is a suit that is calibrated to "approximate the motor, visual, flexibility, dexterity and strength of a person in their midseventies" and is meant to be worn by "students, product developers, designers, engineers, marketing, planners, architects, packaging engineers, and others to better understand the physical challenges associated with aging."

AGEISM: THE LAST SOCIALLY ACCEPTABLE FORM OF DISCRIMINATION

We also have to acknowledge the issue of ageism. According to Ashton Applewhite, anti-aging activist and author of *This Chair Rocks: A Manifesto Against Ageism*, ageism is the last "ism" that is still widely acceptable in our society—unlike racism, sexism, or other forms of bigotry. In Western societies, many people, including older adults themselves, hold negative opinions about adults of advanced age and aging (Kerrie et al., 2001). I can't tell you how many times I've heard older adults say they will never set foot in a senior living community or a seniors' club because they don't want to be around "old people." Oprah Winfrey wrote in her book, *What I Know for Sure*, that "we live in a youth-obsessed culture that is constantly trying to tell us that if we are not young, and we're not glowing, and we're not hot, that we don't matter."

People tend to think of older adults as this monolithic group of people, when in fact, people over the age of sixty-five represent a diverse group of people. In an article written in the American Psychological Association in 2020, Dr. Manfred Diehl, director of the Adult Development and Aging Project at Colorado State University, was quoted saying that older adults are "the most diverse age group." From the generational perspective, older adults today span three generations: the Greatest Generation (born 1901–27), the Silent Generation (born 1928–45) and Baby Boomers (born 1946–64). No one in their right mind would think to place fifteen-year-old Gen Zers in the same "bucket" with thirty-five-year-old Millennials; it's quite obvious to us that these are different age groups with different wants and needs. So why would we place someone who's sixty-five in the same "bucket" with someone aged eighty-five?

Many natural, age-related changes happen in our bodies as we grow older. You can easily tell a boy apart from a man by their height, amount of facial hair, and pitch of voice. Similarly, you can easily tell when this man becomes an old man: gray highlights in his hair and some wrinkles in his skin. What you can't see are the deeper physical and cognitive changes that happen to aging human bodies—ones that aren't visible to the naked eye. How well do your organs function? How fast is your immune response? How good is your memory? How long are your telomeres? Scientists around the world are trying to create an index for biological aging based on known biomarkers for aging (Ferrucci et al., 2019). However, this quest may take a while. In the meantime, we still need to segment this population.

While this can be accomplished in a number of different ways, it is useful to know that researchers like to divide the older population into three age groups:

1. The young-old: from sixty-five to seventy-four years of age

2. The middle-old: from seventy-five to eighty-four years of age
3. The oldest-old: eighty-five years of age and over
   (Lee et al., 2018).

The World Health Organization (WHO) uses eighty years of age as a cutoff between "regular" older adults and the oldest-old (Kalache & Lunenfeld, 1999). It might also be helpful to use a generational lens, especially if you're looking to segmentize older adults as consumers. Today's "young-old" are Baby Boomers, today's "middle-old" are all part of the Silent Generation, and the "oldest-old" are a mix of the Silent Generation and the Greatest Generation.

For many people who don't have frequent daily interactions with older adults, their only examples of old people are their parents or grandparents. When you have such a small sample, it is easy to be influenced by media portrayal of older adults, which is biased against them: negative portrayal of adults fifty-plus is more common than negative portrayal of younger adults (Thayer & Skufca, 2019). While older adults are generally missing from prime-time television, when you do see them on screen, they will often be portrayed as "in poor shape financially as well as physically, sexually dormant, close-minded and inefficient" (ESRI, 2021). With television being "the source of the most broadly shared images and messages in history," it is easy to understand how ageist stereotypes are formed (Gerbner, 2000).

It wasn't always like this. In fact, the definition of the word "elder" in the *Cambridge Dictionary*—"an older person, especially one with a respected position in society"— specifically mentions "elder" refers to an older person who holds a respected position in society. When we examine the cultural differences in people's perception of aging and older adults, we find that more affluent and industrialized societies tend to have less favorable attitudes toward aging and a lower societal status for older adults. Modernization also "devalues their experience-based knowledge, breaks up traditional extended families through urbanization, and shifts control over the means of production from family elders to industrial entities" (Löckenhoff et al., 2004).

There could also be a difference between Eastern/Asian and Western cultures. Some studies found that in some Asian cultures, positive views of aging and high esteem for older adults are more common than in Western, youth-obsessed societies. However, empirical evidence for those East-West differences depends on "how much their measures of aging attitudes emphasized wisdom versus general societal views" (Löckenhoff et al., 2004).

Löckenhoff's 2004 study on the perceptions of aging across twenty-six cultures from six continents surveyed thousands of participants and found participants from Asian cultures held more positive societal views on aging (than Western participants).

"Old age is a ceremony of losses."

DONALD HALL, POET

The older you get, the more losses you accumulate. Upon retirement, you might lose a major way in which you used to define yourself: what you do for a living. Financially, retirement moves you from the accumulation phase to the decumulation phase. Your social circle dwindles: some of your friends and loved ones die, and if you live a very long life, it's possible you will outlive your friends, siblings, your spouse, and even some of your children. You may also lose physical and cognitive abilities. Despite all this, aging is not as bleak as you might think. A 2019 study from the British Office of National Statistics found that **sixty-five to seventy-nine-year-old people are the happiest age group**, and younger people are the least satisfied with their lives.

THE DEMOGRAPHIC SHIFT

Globally, by 2030, more people will be over the age of sixty than under the age of ten (Milken Institute, 2021). The caregiver support ratio for every older adult over the age of eighty is declining rapidly in many Western countries. In the United States, for example, this ratio was seven to one in 2010 but is expected to drop to four to one by 2030 (Riberio et al., 2019).

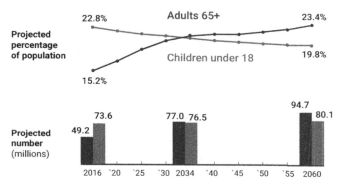

For the First Time in U.S. History Older Adults Are Projected to Outnumber Children by 2034

Note: 2016 data are estimates not projections.
Source: National Population Projections, 2017

Image source: Vespa, Jonathan. "The US Joins Other Countries with Large Aging Populations," United States Census Bureau, March 13, 2018. Accessed September 22, 2021. https://www.census.gov/library/stories/2018/03/graying-america.html.

In the past, multigenerational households were the norm and those who were lucky enough to reach old age lived with their adult children, which meant taking care of one's older loved ones was part of everyday life, just like childcare (Ruggles, 2003). When *Pew Research Center* reported in 2010 that 16 percent of US households are multigenerational, however, it was considered a trend reversal rather than the norm.

Time is precious when you or your loved ones are approaching the end of life, and the COVID-19 pandemic has made that crystal clear. Many families were separated for months because of lockdowns, and many people died in isolation wards without their loved ones by their side. In April 2021, after nursing home residents were vaccinated for the COVID-19 virus and families were finally allowed to go in, *The New York Times* sent photographers to several nursing homes across the US to document the first meeting between older adults and their families.

For many, this was the first meeting in over a year of only seeing each other on video chat or through a window. Carolyn Austin-Tucker, sixty-two, who had not seen her eighty-two-year-old mother in more than a year until they were approved for a visit in April of 2021, said: "I was a little saddened because I realized the year that we missed. Aging is a process, and every moment is precious. The whole reality set in. We missed all of that time. We were happy to see each other, but it was bittersweet."

NOTABLE THINKERS TALK ABOUT AGING
Maya Angelou in *Letter to My Daughter:* "Most people do not grow up. . . . We marry and dare to have children and call that growing up. I think what we do is mostly grow old. We carry accumulation of years in our bodies, and on our faces, but generally our real selves, the children inside, are innocent and shy as magnolias." The reason I love this quote is because Angelou puts so eloquently what many people feel inside: the fact that their chronological age is advanced and their body is showing some signs of aging doesn't mean they are fundamentally different than the child they once were.

As an entertainer with a career spanning more than half a century, you would expect ninety-nine-year-old Betty White sees the fun side of aging. In an interview for *People* magazine in 2013, White shares a similar sentiment to Maya Angelou, saying that the "best thing about being in your nineties is you're spoiled rotten. Everybody spoils you like mad and they treat you with such respect because you're old. Little do they know, you haven't changed."

The Longevity Explorers are group made up of older adults (in their sixties, seventies, eighties, and nineties). It's a unique sharing, evaluation, and ideation community that, among other things, works with tech companies so they can build tech-enabled products and services that "fit" older adults. On their website, I found the following quote:

"We plan to stay in touch and engage with changes as they take place in the world around us. We don't want to live in our own separate 'old people's' world."

(SOURCE: THE LONGEVITY EXPLORERS,

TECH ENHANCED LIFE)

Jane Caro, a feminist social commentator, writer, and lecturer based in Australia, gave a wonderful TED talk titled "Growing Old: The Unbearable Lightness of Ageing." In her talk, Caro explains, in detail, some of the benefits that come with growing old, I'm not going to mention them here because I do want you to look it up and watch it—it's as educating as it is entertaining. One of the points Caro makes is the world wants us to hate and deny this stage of our life, and she feels she is pressured into spending the money she has spent a lifetime earning on anti-aging products. "Why would I want to pretend that I am not my age?" she wonders, continuing on to say that "the loss of your physical beauty teaches you how to become much more aware of the other beauties that you have; that's what that loss does. It enables you to start to understand so much more, what else you have to offer." Caro believes that aging as a stage of life has a lot to teach us, such as "letting go of things like vanity, what other people think of us, approval . . . When you get older, it concentrates the mind, you realize how much shorter the time is that you have left, and that is wonderfully energizing, that makes you move like nothing else." She finished the talk, which was given at TEDxSouthBank just a few days after her friend Jeff Truman passed away unexpectedly at the age of fifty-seven with the following words:

"Get old. You're lucky if you do."

JANE CARO, "GROWING OLD: THE UNBEARABLE
LIGHTNESS OF AGEING" TEDXSOUTHBANK, 2015

I believe this last quote sums it up better than anything. It's so easy to forget getting old is a privilege not everyone gets. With this in mind, we need to appreciate this part of life and make sure we really do respect our elders, which includes making sure we're not excluding them by making digital products and services that aren't age inclusive—more on that in the following chapters.

CHAPTER 2:

Why Does the World Need Agetech?

———

Our world is facing a new set of challenges as the population ages; providing health care to a growing number of older adults who live longer and spend more years dealing with comorbidities, providing care at home to those who prefer to age in place, and making sure people's wealth span matches their life span are just a few of them.[1]

The spending power of consumers over the age of sixty was expected to reach $15 trillion (US) by 2020, making it one of the largest economies in the world, right up there with the United States and China (Nahal & Ma, 2014). However, despite an increasingly growing number of start-ups developing tech-based products and services for this demographic and big tech companies who are also starting to dip their toes into this space, most of the consumer-facing tech being

———

1   The ability to continue living in one's chosen home rather than move into residential care (Davey et al., 2004)

created today isn't designed with the needs and wants of older adults in mind.

Don Norman, born in 1935 and author of multiple books including the highly influential *The Design of Everyday Things*, was Apple's user experience architect from 1993 to 1996. Norman wrote an article for *Fast Company* in 2015 titled "How Apple is Giving Design a Bad Name." In this article, he states that "the new generation of software has made gigantic leaps forward in attractiveness and computational power while simultaneously getting harder for people to use." In another *Fast Company* article from 2019, he makes an observation that "despite our increasing numbers, the world seems to be designed against the elderly." Norman and I met to discuss these points during the time of writing this book—more on this in Chapter Seven.

Why should we design products for older adults? After all, if they are adopting tech in growing numbers, won't they all eventually become tech savvy?

I was born in the late 1980s, so I am what you call a "Millennial." We're considered to be the first digital native generation. Having grown up with technology, it's so intertwined in my life that I can hardly remember what it was like before the world was at my fingertips. I've got excellent digital literacy and I cofounded a tech company, but will I still need special tech to be designed for me when I'm eighty or ninety years old? I believe the answer to that question is yes—I'll explain why I believe that is the case.

Aging in place is arguably the most important concept in aging today. It is a term used to describe the preference of living in one's own home rather than move into residential care. The vast majority of older adults prefer to age in place—to live in the community they know and love, and in their own familiar home, rather than move into senior living. However, for those who require help with activities of daily living (ADL)—such as bathing, toileting, and eating—receiving that care in their home of choice often requires family or paid caregivers, and with a growing care gap and a massive caregiver shortage that is only expected to grow, providing said care becomes a major challenge.

Caregivers are at the top of the most in-demand jobs list according to Indeed in 2020, yet caregiving is a low-paying job that is both physically and emotionally demanding. According to ZipRecruiter, the annual average salary for a caregiver in the United States, in 2021 is $25,878, with an hourly wage that's just above the minimum.

The main challenge that will need to be addressed in the upcoming decades is the caregiver shortage. In his 2017 **book *Who Will Care for Us?* Paul Osterman calculates that by 2030, the US will face "a shortfall of over hundreds of thousands of direct care workers and several million unpaid family caregivers."** Family caregivers and paid caregivers are the ones who help older adults who require assistance with personal care and other ADLs. Caregivers help with essential tasks that will allow aging adults to age in place. The numbers in the US provide an example for the scale of the shortage in the rest of the developed world. These numbers are only expected to grow.

With the increase in demand for caregivers, the cost of care is slowly rising. Genworth, who conducts and publishes an annual Cost of Care Survey, estimated in 2020 that the monthly median cost for in-home care is expected to increase by approximately 30 percent by 2030. There will be a similar increase in the cost of assisted living and nursing homes, which are higher to begin with. Currently, the cost of just two years of in-home care or assisted living could easily surpass $100,000, and many people require care for much longer than that.

You would expect that in this day and age, we could be replacing some of the manual labor caregivers are currently doing, like household chores, with technology. We're not quite there yet. I, however, am optimistic. In recent years, we've seen more and more tech companies tackle the challenges of aging, some more successfully than others. I believe we have a unique, once-in-a-century opportunity within the next ten years to build better technology that will serve the needs and wants of the aging population and ultimately make us a more age-inclusive society.

WHAT ARE THE BARRIERS TO GETTING US THERE?

Since most if not all the tech that's being used today and that will be developed in the future to tackle the challenges of aging depends on having an internet connection, we first need to bridge the digital divide. The Organisation for Economic Co-operation and Development (OECD) defines the digital divide as "the gap between individuals, households, businesses, and geographic areas at different socio-economic levels with regard to both their opportunities to access information and communication technologies (ICTs) and to their use of the Internet for a wide variety of activities." In short, it is the gap between those who have access to computers and internet, and those who don't (OECD, 2002).

Our society has been going through a digital transformation in recent years, with computers and the internet playing an increasingly growing role in our everyday lives and more and more services that used to be in-person being replaced by apps and websites. Bank branches, for example, are being closed all over the world. According to "The Death of the Banks," a study published by *Self*, the current trend suggests that **all US bank branches could be closed by 2034**. Banking is just one example for an essential service that has been gradually phasing out its real-world presence and switching to a digital-first presence.

The COVID-19 pandemic has accelerated society's digital transformation. According to a 2020 McKinsey survey, "The COVID-19 crisis has accelerated the digitization of customer interactions by several years." This creates new opportunities for remote workers to work anywhere and for people who are homebound to participate in more activities—since more of them are now available online. It also puts a spotlight on the digital divide and the urgency of making sure no one is left behind as the world races forward. In Chapter Eight, we will discuss the role of governments and NGOs in these efforts.

When COVID-19 vaccines were starting to roll out, so were news stories about older adults not being able to get an appointment to get vaccinated simply because they didn't have internet access at home or the digital literacy to register online for an appointment. Getting everyone online isn't just for the sake of convenience, it's a necessity.

For many families, mine included, the family conversation has also moved, or rather, re-emerged online. We use WhatsApp—an instant messaging platform with multimedia capabilities. In our family group chat, we share texts, photos, videos, and voice messages. Occasionally, we will use it for a family-wide video conference call. My maternal grandmother has a smartphone and knows how to use WhatsApp, so we know that if we send messages or pictures, she will see it. My paternal grandmother, on the other hand, does not own a smartphone or a computer (nor does she want to), so if I want to show her a picture that I took when I went to the beach, I have to drive up to her house (two hours away) and show it to her on my phone.

This isn't just my story; when we see the stats, we understand many families all over the world share this experience. When the family's elders don't master the digital tools and are left behind in the digital divide, they can be accidentally excluded from the family conversation. It's not because families deliberately do it; in fact, in most cases, families are an enabler for crossing the digital divide. They will buy a computer or smartphone for their older loved ones, they will set up an internet connection, and they will teach them how to use it. Not everyone has family members who can do that, and since most of the technology we use today is designed by younger adults, for younger adults, we need more than an internet connection to cross the digital divide—we also need digital literacy education.

While society may believe older adults are not interested in tech, they are adopting tech in growing numbers. Two *Pew Research* reports from 2017 and 2021 have shown a steady climb in tech adoption by older adults in recent years, with 67 percent of Americans over the age of sixty-five using the internet in 2017, and 75 percent in 2021. Device ownership continues to climb as well, and according to AARP's "2021 Tech Trends and the 50 Plus" report, the COVID-19 pandemic has been a major accelerator for adoption of video chat technologies and participation of older adults in online events. The report also found that although ownership rates of smart technology devices continue to rise, lack of accessibility and affordability still block many older adults from using technology. Lack of knowledge is another barrier, and 39 percent of survey respondents said "they would use technology more often if they knew how."

Initiatives to help older adults cross the digital divide and learn how to use technology have been around since the 1980s. Mary Furlong founded SeniorNet in 1986 with a mission to help older adults "take advantage of new technology that can improve their quality of life, reduce isolation and engage in new and interesting ways" (SeniorNet, 2021).

Tom Kamber founded Older Adults Technology Services (OATS, now a part of the AARP) in 2004 with a similar mission. For a nonprofit, the resources that were available to fund OATS' mission were mostly donations or payments from foundations and city contracts. This was great seed funding. It helped Kamber and his team lay the foundations of what would eventually become a national operation and establish six Senior Planet centers, physical spaces in which older adults could take multiweek technology courses, learn computer basics, and develop life-enhancing computer skills. With Senior Planet centers in five different states, OATS was able to teach fifty thousand people how to go online.

A few years in, the people at OATS realized older adults wanted to learn online, but they didn't have the means to do it. Their entire program was face to face and every class was taught in a physical space, with a trainer teaching roughly twelve people—a challenging operation to scale. Around that time, Kamber met Steve Ewell, the executive director of the Consumer Technology Association foundation (CTA) at a conference and learned that the CTA foundation was focused on innovation in aging. He proceeded to apply for OATS' first innovation grant and received the funding needed to learn how to digitize their training so they could teach online. In 2016-17, OATS built an online training program, which is now the main way they teach. Fifty percent of OATS curriculum was already digital prior to 2020, so when the COVID-19 pandemic hit and they had to do a digital pivot, they were able to relaunch all their programming online within forty-five days (OATS CES2021 Interview, 2021).

This wasn't the end of OATS' efforts to help more older adults adopt technology. They also realized that internet connectivity, or lack thereof, was a major barrier for adoption and partnered with the Humana Foundation to create the Aging Connected initiative. According to the "Aging Connected" report published by OATS and the Humana foundation in early 2021, **twenty-two million older Americans don't have internet broadband at home.**

When I first got a glimpse of this report in early 2021, I was simply blown away by the fact that 42 percent of older adults in the wealthiest country in the world don't have access to basic twenty-first century infrastructure that is broadband internet in their home.

I wondered what the numbers in senior living communities across the United States were, where Wi-Fi is still considered an amenity, so I reached out to Arthur Bretschneider from Seniorly to inquire about how many senior living communities report having Wi-Fi on their premises. Bretschneider, who is a third-generation senior living operator, founded Seniorly in 2014 with the mission to help older adults find the right senior living community to fit their needs and budget. The company has an extensive database of senior living communities across the US, and according to Erica Powell, senior strategy and analytics manager at Seniorly: "out of all senior living communities that have self-reported their amenities on the platform, 40 percent of small communities (under twenty beds) and 86 percent of large communities have Wi-Fi."

WHAT ARE THE UNDERLYING CAUSES?

A EuroAgeism policy brief published in 2021 by Köttl & Mannheim mentions ageism as a possible cause. And there's also the matter of usability—it is a key component to our ability to use different products, especially tech-based products and services. In general, today's older adults are healthier than previous generations (Czaja and Lee, 2007). However, as people grow older, naturally occurring changes in vision, hearing, motor skills, and cognition can impact their ability to use certain user interfaces.

Don Norman, who's now in his eighties, wrote in his Fast Company article from 2019 that "new technologies tend to rely on display screens, often with tiny lettering, with touch-sensitive areas that are exceedingly difficult to hit as eye-hand coordination declines."

According to the American Optometric Association, people as young as forty years of age can experience difficulty reading in low light and focusing on near objects and can even experience changes in color perception. This impacts people's ability to read text that's written in a small font size and low contrast. However according to Norman, "Companies insist on printing critical instructions in tiny fonts with very low contrast," making labels impossible to read without flashlights and magnifying glasses.

As for hearing, according to a Johns Hopkins Medicine article, "Age-Related Hearing Loss (Presbycusis)," one out of three adults over the age of sixty-five experiences hearing loss. Some of the symptoms may include speech or other sounds becoming muffled or slurred and loss of the ability to hear high-pitched sounds, making female voices harder to hear than male voices, and also making it harder to distinguish between "s" and "th." Conversations are difficult to understand, particularly when there is background noise. Closed captions are a good way to solve this when watching TV or movies, but they aren't available for every video on the internet, and the solutions for captioning live, real-life conversations aren't very well known.

On top of the inevitable physical changes in aging, there's also cognitive decline caused by structural changes to the brain. However, the way this cognitive decline affects people isn't uniform. Abilities like processing speed and memory decline over time, when others, like vocabulary, are more resilient to aging and might even improve with time (Harada et al., 2013). For those who need to learn how to use digital technologies for the first time later in life, these changes in cognition make it harder (Ya-Huei et al., 2019).

Other physical aspects that are affected by aging—such as dexterity, upper limb posture and even skin dryness—can affect one's ability to use touch screens—which have become ubiquitous not only in personal, handheld devices but also in service stations in banks and supermarkets.

All these things have to be taken into consideration when designing technology if said technology is to be age inclusive.

When designing tech for older adults, there's another important element that has to be taken into account—desirability. Will this bring value to people's lives? Will they want to use it? Will it make them feel good about themselves, rather than making them feel old, frail, frustrated, or incompetent?

WHY I'M HOPEFUL

In recent years, I've witnessed a steady increase in the number of tech companies specifically aiming to serve the needs and wants of the aging population, along with the number of funds looking to invest in these types of companies. On top of this, the intersection of two trends makes me hopeful that we're on the right path. One is the rise of Millennial caregivers, and the other is the retirement and aging of Baby Boomers and the enormous amounts of wealth that they hold. Millennials and Baby Boomers are currently the two largest living adult generations (Pew Research Center, 2020).

In 2018, the AARP Public Policy Institute released a report focusing on the 25 percent of family caregivers in the United States that are Millennials, people born between 1980 and 1996. As the first generation of "digital natives," Millennials expect to be served by technology. However, when they assume the role of the family caregiver, they often find there are no good digital solutions to tackle some of the challenges they are facing. Baby Boomers often face a similar situation, whether they are caregivers for a spouse or a parent or are starting to feel the effects of old age.

I expect many of the challenges that caregivers are facing today will eventually be solved by caregivers who couldn't find the right solution for themselves and decide to build a tech-enabled product/service that does. In fact, many of the companies on the agetech market map actually started this way, because of a personal experience the founders had caring for their own parents' grandparents (Etkin, 2021). On the consumer side, these entrepreneurs will find Baby Boomers who are desperately looking for solutions and can afford to pay for them.

WHAT EXACTLY IS AGETECH?

"Gerontechnology is a technology domain that links existing and developing technologies to the aspirations and needs of aging and aged adults" (Bronswijk et al., 2009). The prefix "gero," or "geron," represents old age, whereas technology is defined by Merriam-Webster dictionary as "the use of science in industry, engineering, etc., to invent useful things or to solve problems." It appears the definition of gerontechnology is quite broad and can cover almost any modern technology older adults use to meet their needs and aspirations—even a rollator or a car could fall under this category.

Agetech is a term used to describe **digital** technology that's built around the needs and wants of older adults and those who care for them, preferably while including them in the design process.

This definition expands on the definition of gerontech when it comes to the target audience (or group of users), while narrowing it down when it comes to the type of technology— digital. This means that a care-coordination app, being used by several family members to coordinate who's dropping grandma off at the movies to meet with her friends this week, also falls into the agetech category, along with a smart pillbox that will remind her to take her vitamins.

It is obvious by now that agetech, digital technology that's designed with the needs and wants of older adults in mind (and preferably, in practice, by including them in the design process) is a must-have component for aging in the twenty-first century.

CHAPTER 3:

Why Now?

In early February 2020, as I was waiting at the bus stop to get back home from work, my phone rang. It was a dark winter night, and on the other side of the line was Eran, an Israeli agetech entrepreneur based in New York. We met through a mutual acquaintance, and he offered me a spot to come pitch my start-up at the eCap Summit in Miami. At this point, I had to confess I had never heard of this summit before, but Eran reassured me this was a great opportunity and he wouldn't be offering it to us unless he thought we could make valuable connections there.

Only three start-ups could present, and we were being offered a spot. Airfare alone would cost us a small fortune (we were bootstrapped at the time), but it could be a game changer for a company at our stage. So of course, we said yes. Little did I know this would be the last in-person event I would get to attend for a very long time.

The eCap Summit wasn't a trade show where we would get lost in the crowd among hundreds or thousands of other vendors, dangling shiny objects in the faces of potential buyers. It was where hundreds of nursing home owners and operators met annually to talk business. Since most of the attendees happened to be Orthodox Jews and there were very few women attending, I wasn't quite sure what to expect in terms of networking.

Would we blend into the crowd, or would we stick out like a sore thumb? As it turned out, we stood out big time, but thankfully, no sore thumbs were involved, and standing out in the crowd turned out to be a great advantage in this context. Whenever I would speak to someone, we would exchange business cards and sometimes schedule a follow-up meeting for later that day or the next day. When someone I met wanted to introduce me to someone else he thought would be interested in our product, he could easily locate me in the crowd and make the introduction.

In those days, COVID-19 was just starting to rear its ugly head, but no one seemed to be particularly worried about it, and I figured Miami was just as safe as Tel Aviv, where I'm from. For an entrepreneur, the risk of a few days in quarantine seemed to be worth the shot of landing a big client, which we eventually did.

Who knew this would become a once-in-a-century pandemic that would kill millions of people? When I was preparing to board my flight to Miami, an airline representative had asked me if I had been to China in the fourteen days prior. This was my first clue that things were about to change. The second clue was when I found myself spending the entire seventeen-hour journey back home wearing a surgical mask. When

I got home safely in early March 2020, COVID-19 restrictions had slowly started taking hold in our lives.

The worst part was the uncertainty. How many lives would be lost? How many people would lose their job? Will the economy crash? When my grandparents' senior living community went into lockdown and stopped letting visitors in, I had to ask myself: Will I ever see them again?

In Israel, independent living is called "Diyur Mugan," which roughly translates to "protected housing." My grandparents were undoubtedly protected, but they were also isolated from us and the outside world, which quickly became a dangerous place for older adults.

Thankfully, they had Wi-Fi in their apartment. My grandmother had a smartphone and knew how to use WhatsApp for messaging and video chat, and my grandfather knew how to use the computer for his online banking and bill paying needs. For Passover, which is one of the major Jewish holidays, we got them a tablet. The whole family was scattered for the first time, so we had Passover dinner over Zoom. My grandmother later told me she thought it was the most wonderful Passover dinner we'd ever had.

My grandparents were far better off than many other older adults. They had each other, they had a balcony overlooking a park, and they had the means to connect with us. I really don't know what we would have done otherwise.

For contrast, my paternal grandmother, who doesn't have a computer or a smartphone, was significantly worse off. Even if we got her devices and set up an internet connection in her apartment when it became clear that lockdown was going to take months, she wouldn't know how to use them. Who would have thought that she would have to prepare for a global pandemic?

When I came to visit her (from afar) on that same Passover, only to say hello to her when she was standing on her first-floor balcony, she cried.

It took almost a year for me to be able to set foot in either of my grandparents' apartments without fear that I was putting their health at risk. As a family, we only allowed ourselves to visit them indoors when all of us were fully vaccinated, wearing masks, and keeping the windows open.

The COVID-19 pandemic brought to light some of the challenges older adults have been dealing with for years. The digital divide did not suddenly appear in March of 2020, but suddenly everyone was made aware of the fact that despite all the wonderful progress tech had brought into our lives, we were doing a disservice to our elders when many of the tech solutions that are so common today exclude them. "The struggle to get vaccines when they were first available nationwide demonstrated the difficulties in a digital world. Nearly every state has required citizens who are eligible to book appointments online, but tension emerged as some elderly Americans don't know how to use a computer or have access to one" (Bono, 2021).

I believe we are in a pivotal moment in human history. The intersection of two trends: the aging of the population and a fourth Industrial Revolution. "The digital revolution that has been occurring since the middle of the last century. It is characterized by a fusion of technologies that is blurring the lines between the physical, digital, and biological spheres" (Schwab, 2016). This provides us with a unique opportunity to reshape our views on aging, how technology serves us and our elders, and how we view and use technology.

The third Industrial Revolution, which started in the 1950s with the rise of electronics, telecommunications, and computers, has fundamentally changed the way we live, work, and communicate with one another. As a result, many more people were able to become knowledge workers and earn a living working with their brains rather than by doing manual labor. It enabled the creation of tech hubs like Silicon Valley, which created technology that could be used by billions of people all over the world, and it also allowed our world to become a global village in which physical distance is almost irrelevant and people from different countries and cultures can congregate in one place—the internet—to create friendships and do business. Everything is faster in this third revolution.

In a few short decades, fax machines replaced the postal service and telegrams as a means of business communication, only to be replaced by emails, which became the standard means of communication.

From the time of the first Industrial Revolution, businesses were incentivized to adopt new technologies to reduce the dependency on manual labor and increase profitability. In recent years, big corporations as well as small and medium businesses became obsessed with "digital transformation"—"the process of using digital technologies to create new—or modify existing—business processes, culture, and customer experiences to meet changing business and market requirements" (Salesforce, 2021). For some, digital transformation is do or die. Remember Blockbuster? It used to be one of the largest providers of in-home entertainment. However, the company couldn't transform itself into an all-digital company to compete in a new, internet-based world and filed for bankruptcy in 2010.

HOW DOES ALL THIS RELATE TO INDIVIDUALS? MORE SPECIFICALLY, TO OLDER INDIVIDUALS?

First of all, it's important to understand that many practices from the business world, like sending emails, trickle into our personal, everyday lives. Email is just one example. It started in the academic/business world and later became a ubiquitous means of communication. Zoom calls were primarily used for business calls up until the COVID-19 pandemic, and now Zoom has become a household name people use to connect with friends and host events during lockdowns.

Then there are touch points between consumers and consumer-facing businesses. They usually interact with consumers when they're making a purchase or when they need customer or technical support. The reformation of customer service is driven by consumers who make purchases online and expect to receive service on multiple channels and devices (Deloitte, 2013). In a highly competitive business environment, businesses strive to keep up with consumer demand and "digitally transform" the way they do business.

Eneda Bourne, seventy-five, a retired teacher from South Carolina, said in a 2021 interview for Sal Bono from *Inside Edition* that technology sometimes feels like a high-speed train. "A train is coming into the station slowly, you get on there, it's not coming to a full stop. You hop on that train, and when you get on that train, you put on your belt, your seatbelt, whatever. . . . And if you can stay on that train, the joy at the end of that train is fabulous. But some people can't make the journey. It's too difficult. So, with technology moving so fast, so quickly, you have to hang in there and give it your all. Don't let go. Because if you let go, you might as well get off the train. And if you get off the train, you are out of the loop with life, and with society, and with your grandchildren, and with your adult children."

When I spoke with George Lorenzo, a sixty-seven-year-old internet entrepreneur who established three websites, including one that's called Old Anima (which translates to "old soul"), he had a similar feeling. "I feel like because I'm not a digital native, I'm kind of behind the eight ball in some ways, even though you think I'm an older adult who knows more about it than most. I don't feel that I do. I feel like I need to learn more all the time. Most older adults don't have to keep up with it in the same manner that I do. . . . Today, you have to have foundational digital skills. And that means you have to at least know how to email properly, you know, and how to communicate online, and perhaps go into a Zoom meeting with your colleagues and all that stuff. So, I'm sure a lot of older adults have a much more difficult time than I do."

In the early 2000s, personal computers and home internet became ubiquitous, and it became possible to gradually replace phone and face-to-face communication between consumers and service providers with written communication (via chat or email). The telephone was replaced by emails, and the cashier at the grocery store was replaced by a checkout button on the store's website. For businesses, digital transformation allows them to cut down on costs. Having a physical store, or multiple physical stores, is much more expensive than having one e-commerce website and one (or several) warehouses and fulfillment centers. The proliferation of online retailers like Amazon, who didn't have brick-and-mortar stores, and was said to be putting brick-and-mortar bookstores out of business, is a prime example for that. Many companies, including some of the largest tech companies in the world, which serve billions of end users, don't provide customer support over the phone.

Almost anything can be done online these days: everyday activities like shopping, banking, bill-paying, reading the news, and consuming entertainment. Working in many "knowledge" professions can certainly be done online. Your laptop can become your window to the world; you can use it to make a living, entertain yourself, and have food delivered to your door. You can even use it to participate in online fitness classes. You never have to leave your house again if you don't want to. For some people, especially people with physical disabilities, having the world at their fingertips is progress that makes life easier.

However, progress and digital transformation also mean all consumers either have to adapt to these changes or fall behind.

When the COVID-19 vaccine distribution began in early 2021, most countries set up online systems for booking vaccine appointments. This resulted in many older adults being unable to get the vaccine without outside assistance.

Jeremy Novich, a clinical psychologist in New York City who has started a group to help people navigate the technology to get a vaccine appointment, told *NPR*'s Will Stone in 2021, "You can't have the vaccine distribution be a race between elderly people typing and younger people typing. That's not a race, that's just cruel." When face-to-face or phone service is no longer the default option, what can we do to make sure we're not leaving less digitally savvy people, like some of our elders, behind?

There are many other basic services and customer-facing business practices that are expected to go all-digital in the next few years. The cashless economy is on the rise. What started with credit and debit cards decades ago is now moving toward mobile payments. In fact, some brick-and-mortar businesses are going cashless altogether, which helps them cut down on operating costs and be more environmentally friendly, since they don't have to transport cash to and from the bank. This could also potentially protect them from robbery, since there is no cash in the cash register. Paper manuals are mostly a thing of the past, and if you purchase a new product, you'll have to look for the manual online.

Some very basic services have increasingly gone online in recent years. Bank branches are closing rapidly in favor of online banking. Post offices are closing because people aren't sending letters anymore—they exchange emails.

Larry Irving, cofounder of The Mobile Alliance for Global Good, wrote to *Pew Research Center*: "There is almost no area in which digital technology has not impacted me and my family's life. I work more from home and have more flexibility and a global client base because of digital technology. I monitor my health and keep my physician informed using data technology. My wife has gone back to a graduate school program and is much more connected to school because of technology. My entertainment and reading options have exploded exponentially because of new technologies. Use of home speakers, Internet of Things, AI [artificial intelligence] and other emerging technologies is just impacting my life and likely will become more central."

In 2015, *Foreign Affairs* published an article about the fourth Industrial Revolution that opened with the following words: "We stand on the brink of a technological revolution that will fundamentally alter the way we live, work, and relate to one another." The COVID-19 pandemic that threw the world into turmoil in 2020 accelerated many of the digital trends that started a few years ago. "In just a few months' time, the COVID-19 crisis has brought about years of change in the way companies in all sectors and regions do business" (McKinsey, 2020). It forced even the most traditional organizations—like governments and health care organizations, which aren't exactly known for being agile in adapting to change and adopting innovation, to change the way they operate. It also made it abundantly clear that when it comes to older adults, digital inclusion is no longer a luxury—it's a necessity. "Seniors isolated by lack of access to reliable technology face personal hardships including physical and mental health issues that affect the quality and duration of their lives" (OATS and the Humana Foundation, 2021).

Including older adults in the design and development of new tech is not only the right thing to do, it also makes a lot of sense business wise. The global spending power of people over the age of sixty was expected to reach $15 trillion in 2020 (Nahal & Ma, 2014). According to the AARP's Longevity Economy Outlook report, Americans over the age of fifty are expected to be spending more than $200 billion a year on technology by 2030. The COVID-19 pandemic accelerated this trend and the AARP's 2021 Tech Trends and 50+ report said the pandemic is driving an exponential increase in tech spending. This research also found that, compared to 2019, tech spending among older adults in 2020 went up 194 percent, from $394 to $1,144.

BABY BOOMERS REACHING OLD AGE

By 2030, each and every Baby Boomer on the face of the earth will have passed the age of sixty-five. The Baby Boomer generation was, and still is, extremely influential in our society—not only culturally, but also financially. In the United States, Baby Boomers control over 50 percent of the country's wealth, compared to 5 percent that's currently in the hands of Millennials (Adamczyk, 2020). Members of this generation invented the internet as we know it and were responsible for the first commercially successful personal computers. According to the World Wide Web Foundation, the World Wide Web was created by Sir Tim Berners-Lee, a British scientist born in 1955, during his work at the European Organization for Nuclear Research (a.k.a. CERN). You may have also heard of Steve Jobs and Steve Wozniak—two Boomers who founded Apple Inc., the company that made the first commercially successful personal computer (Britannica, 2021). Despite the fact that they're not considered to be digital natives, Baby Boomers use technology more than previous generations (Pew Research Center, 2019). It's getting harder and harder to find a Boomer who doesn't know their way around technology in developed countries.

Because Baby Boomers have discretionary income and they won't settle for tech that doesn't bring value to their life or make them feel bad about themselves, they're a force to be reckoned with in the longevity economy. Since they are more tech savvy than their parents, they expect service providers to be accessible through digital means, and to have twenty-first century infrastructure. When Baby Boomers inquire about a senior living community for their parents or themselves, they might present new demands—like high-speed internet throughout the premises. Chip Burns, CIO of Sun Health Communities, said in an interview to Tim Kridel from *HealthTech Magazine* in 2019 that "these new seniors want to know that we are on top of technology—even if they're not. . . . Their families want to know that mom and dad are going to a place where they will not only have the technology themselves but the family can keep in touch too."

THE RAMIFICATIONS OF THE COVID-19 PANDEMIC.

The global COVID-19 pandemic that erupted in 2020 has been rough on humanity. Many lives were lost, and many people's lives were thrown into disarray. The vaccine rollout has enabled some parts of this world to reclaim some sense of normalcy, but it is obvious by now we're not going back to the way things were anytime soon, if ever. The pandemic brought into public awareness the digital divide and the fact that no one can manage without tech. It also brought awareness to challenges older adults have been dealing with for years, like loneliness and social isolation. Until recently, these issues were only spoken about in circles like care professionals and among older adults themselves, but suddenly, they began to make headlines.

In March 2021, Scott Galloway wrote in his *Medium* article "This Is the Best Time to Start a Business" that historically, post-crisis periods are extremely productive eras. I believe this is an opportunity. The spotlight that has been turned onto the challenges of aging during this pandemic, the fact that suddenly everyone realizes that having a digital divide between generations is not acceptable, and the momentum that's being generated by millions of older adults adopting cutting edge technology at the same time is a huge opportunity for entrepreneurs. Now is the time to seize it.

CHAPTER 4:

The Challenges of Aging

———

I grew up wanting to be a scientist. Specifically, I wanted to become a biologist and help find a cure for cancer or HIV. But in 2010, an ad on the students' association website caught my attention, and the events that followed changed my career trajectory and my view of the world forever.

My fascination with science and technology and my love for gadgets have been a part of me for as long as I can remember. Being an "elder Millennial" means that although I am a digital native, I can still remember what it was like before we had internet at home, and if I wanted to learn something new, all I could do was read a book or ask a grown-up. After I read all the science books I could find at home (most of them were probably outdated since neither of my parents were scientists) my parents went on to buy me some age-appropriate science books and supported me when I decided to study biology in high school, and later on, as I enrolled at university to pursue a BS in life sciences.

I was on the path of pursuing my passion and becoming a real scientist! Seeing old pictures of myself wearing a lab coat and holding a petri dish with fragrant bacterial contents still puts a smile on my face.

While at university, I discovered the universe had other plans for me. I saw an ad looking for volunteers to document the life stories of Holocaust survivors. My late grandfather was a Holocaust survivor. Unlike some survivors, he was very open about his life story and was ready to give testimony wherever, whenever. As a family, we were lucky to document his story in a book shortly before he suffered a stroke and lost the ability to communicate effectively.

When I saw that ad, I was compelled to apply, and soon after that, I found myself a part of a student organization that was not only documenting the life stories of Holocaust survivors, but also helped them apply for government grants and stipends. By that time, Holocaust survivors were all over the age of sixty-five, so the challenges they were dealing with weren't just related to the fact that they went through a horrific life experience in early childhood, but also related to the fact that they were growing older. I spent two years volunteering with and managing other volunteers in helping Holocaust survivors. I remember the first time I stood in a classroom full of volunteers that were about to get trained and then go out to volunteer, understanding I was responsible for getting them there and feeling grateful to have this opportunity to make an impact on people's lives.

During those two years I had realized how much of an impact one person could make if they were in the right place at the right time, and while I did graduate from university with a BS in life sciences, those two years left their mark on me, and I never set foot in a laboratory after graduation. Instead, I embarked on a career in the nonprofit sector working with Holocaust survivors and eventually moved on to working with local communities of older adults in Israel and getting my master's degree gerontology.

Working with older adults in my twenties has definitely helped shape my worldview. When I'm out and about in the world, I often think about how an older person might experience what I am experiencing. Whether I'm crossing a busy intersection in which the duration of the green light that's meant for pedestrians isn't long enough, or I'm using a digital service that's clearly designed for digital natives, it makes me angry. These are just two examples for the way our world isn't designed for older adults.

People face many new challenges as they get older. They can be referred to as "the challenges of aging."

Social isolation and loneliness are two of these challenges. Despite all the wonderful work that nonprofit organizations are doing to serve the needs of older adults, most of them have limited resources and therefore limited reach. When I was working/volunteering in the nonprofit sector, I caught a glimpse of this. There were many times older people (especially older women) confided in me and told me they were lonely. It was heartbreaking, and there wasn't much I could do except pair them up with a volunteer or two, or in some cases seven—one for each day of the week. But even those who were fortunate enough to have a volunteer come over to visit them every day still spent most of their waking hours at home, by themselves.

One thing that is important to know about Israel is that it's a small, densely populated country, with close-knit families and a Mediterranean culture. It's quite common for families to live within a short driving distance apart, meet every week for a multigenerational Friday dinner or even spend an entire "Shabbat" (at least twenty-five hours straight) together on a regular basis. This doesn't mean older adults in Israel don't get lonely, but those of them who have close families will spend less time alone. Despite this, Israel isn't so different from the rest of the world when it comes to loneliness among older adults. Our elders feel lonely too.

When I got the call to join Intuition Robotics and work on a social robot that will help alleviate loneliness for older adults, I realized this could be a game changer. Despite our best efforts in the nonprofit sector, what we were doing wasn't scalable, and technology is. When I joined the company, I was tasked with figuring out what exactly older adults wanted us to build for them. At the end of this journey, ElliQ was born.

Being curious by nature, once I took the job, I became obsessed with learning as much as I could about tech for older adults. I read every academic paper I could get my hands on about social robots for older adults and tech adoption by older adults and spoke with countless older adults about their use of technology. I also spoke to family caregivers about the challenges of caring for an older loved one, especially someone who lives far away. At some point, I learned there was a global network of aging innovators called Aging2.o, which had a chapter in Tel Aviv. One day, I found myself in the brightly lit Google for Startups campus, in a room full of innovators who were all passionate about the same thing I was: tackling the challenges of aging and improving the lives of older adults with technology.

SO WHAT ARE THE CHALLENGES OF AGING?

Several organizations have taken it upon themselves to define the different challenges the aging population is facing through the prism of innovation priorities and opportunities. Aging2.0 was probably the first. This international organization was founded in 2012 with a mission to accelerate innovation to address the biggest challenges and opportunities in aging. Since then, its community has grown to over forty thousand innovators (older adults, senior care providers, thought leaders, and entrepreneurs) across thirty-one countries and 120 cities, and it is currently the largest global organization that promotes innovation in aging.

The dialogue with stakeholders from across Aging2.0's network resulted in eight "Grand Challenges" that are the innovation priorities for the Aging2.0 community. They include:

- Engagement & purpose
- Financial wellness
- Mobility & movement
- Daily living & lifestyle
- Caregiving
- Care coordination
- Brain health
- End of life

Image source: Aging2.0

Another organization that took a shot at defining the challenges of aging is AGE-WELL, Canada's technology and aging network. The name AGE-WELL is an acronym for Aging Gracefully across Environments using Technology to Support Wellness, Engagement, and Long Life. This organization is dedicated to the creation of technologies and services that benefit older adults and caregivers and help older Canadians maintain their independence, health, and quality of life through technologies and services that increase their safety and security, support their independent living, and enhance their social participation. AGE-WELL defined the following challenge areas:

- Supportive Homes & Communities
- Health Care & Health Service Delivery
- Autonomy & Independence
- Cognitive Health & Dementia
- Mobility & Transportation
- Healthy Lifestyles & Wellness
- Staying Connected
- Financial Wellness & Employment

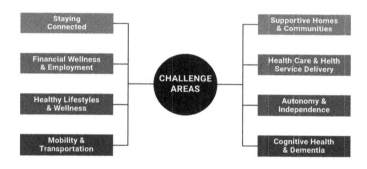

AGE-WELL Challenge Areas:

Image source: TheGerontechnologist.com. Data source: AGE-WELL

Last but not least, the United States National Science and Technology Council assembled a task force on research and development for technology to support aging adults. In 2019, they published a report in which they described, in detail, the six functional capabilities that are critical for maintaining independence as people grow older, and in which technology could have a positive impact. Those six were then broken down to two to three sub-functionalities each.

- Key activities of independent living that include hygiene, nutrition, and medication
- Cognition that includes cognitive training, cognitive monitoring, and financial security
- Communication and social connectivity, which includes hearing, communication with diverse communities, and social communication technologies
- Personal mobility, which includes assisted movement, rehabilitation, and monitoring and safety
- Transportation, which includes driving and public transportation
- Access to health care, which includes telehealth and eCare Planning

It is not surprising there are common themes to the challenges brought to light by these three different organizations. Despite the different wording, these six common themes emerge:

- Finance
- Physical health
- Cognitive Health
- Communication & Social Connectivity
- Mobility & Transportation
- Activities of Daily Living

I would argue aging in place should be the prism through which we look at all these challenges, because it's universally accepted as the most desirable way of living for most older adults and because it's impacted by most, if not all of the above challenges.

Image Source: TheGerontechnologist.com.
Data Source: Task Force on Research and Development for Technology to Support Aging Adults Committee on Technology of the National Science & Technology Council. "Emerging Technologies to Support an Aging Population." Trump White House Archives. March, 2019. https://trumpwhitehouse.archives.gov/wp-content/uploads/2019/03/Emerging-Tech-to-Support-Aging-2019.pdf.

There are also additional challenges that I believe we should now start looking at: employment in later life, leisure, and self-fulfillment. There's also caregiving, which Aging2.0 did include in the Grand Challenges.

If we want to look at this from a broader lens and look at the challenges that the aging of the world's population poses for society as a whole, the working paper "Population Aging: Facts, Challenges, and Responses," describes the following challenges:

- The size and quality of the workforce
- Noncommunicable diseases
- Financial issues
  (Bloom et al., 2011).

Let's take a closer look at some of the challenges:

HEALTH & WELLNESS

Health is definitely a top priority for people as they grow older. Good health enables you to stay independent and keep living your best life. The Global Wellness Institute defines "wellness" as **the active pursuit of activities, choices and lifestyles that lead to a state of holistic health."**

The health care industry is one of the largest industries in the world. The global health care industry was worth $8.45 trillion in 2018, and global health care spending could reach over $10 trillion by 2022. "Aging and growing populations, higher rates of chronic health conditions, and exponential but costly advances in digital technologies will continue to push global healthcare expenditures upward" (Stasha, 2021).

Older adults are the biggest consumers of health care services—people aged fifty-five and over account for over half of total health spending (Sawyer and Claxon, 2019). This is probably the reason many people mistakenly assume the agetech ecosystem contains mostly digital health products. While digital health and wellness are major subcategories in the agetech ecosystem, they're not all it includes.

But since health is so hugely important for older adults and those who care for them, many solutions exist for tackling the major challenges, like falls and medication management.

In fact, fall detection (and prevention) is one of the most saturated categories in the agetech space, and although there isn't one solution that dominates the space, there are plenty of vendors developing wearables and home sensors for this purpose. Some of them are nicely designed and are available for use by the general population, like the Apple Watch or the gait assessment feature in iOS. It's not uncommon for these types of fall detection systems/wearables to use artificial intelligence to detect subtle changes in movement patterns that could indicate an increased risk of falling and send out an alert that can trigger professional intervention.

There are also online fitness programs designed to help older adults gain muscle mass and work on balance and mobility with the purpose of preventing falls.

Amanda Rees is an entrepreneur and a former family caregiver. When Rees was a student at Princeton, she studied chemical and biological engineering and earned a certificate in dance. Soon after graduating, she decided to move in with her grandmother and spent her early twenties living with her in the San Francisco Bay Area. Little did she know that this decision would become a defining moment that would change the trajectory of her career.

About eight years into co-living with her grandmother came the challenges of caregiving. Rees' grandmother would frequent the hospital and started experiencing something that is quite common as people get older—falls. According to the Centers for Disease Control and Prevention (CDC), "One out of four older adults will fall each year in the United States, making falls a public health concern, particularly among the aging population. About thirty-six million older adults fall each year—resulting in more than thirty-two thousand deaths."

As a young adult living right in the middle of Silicon Valley, Rees started looking for solutions, and all she could find were sensors to monitor her grandmother and detect if and when she does fall. "Day-to-day, it really fell on me to figure out what I can do to help her stop falling so often."

Rees started looking into the research about falls in older adults and how to prevent them, discovering that "most falls are preventable. If you can build back the strength, balance and the stability." And so, thanks to her background in dance, she was able to start working with her grandmother on those elements, in the hope that it would prevent her from falling again in the future. She soon began working with groups of older adults in the Bay Area: "I felt like there was something really powerful around adapting exercise programs, making them fun and engaging and building someone's strength and balance back."

Rees teamed up with life partner, Hari Arul, who has a background in health care software and technology, to build a product that would be "as accessible and available to people as possible, by partnering with doctors and health plans so that it doesn't just fall to them to find it and pay for it directly" And so, Bold was born—a digital exercise platform that provides science-backed personalized workout plans that improve balance, strength, and mobility. Now, older adults from all over the world can log into Bold's website, go through an assessment and get a custom training program that fits their needs and preferences.

During the COVID-19 crisis, when gyms and fitness studios had to close their doors for months, the need for a product like Bold became even more apparent. As people grow older, they experience what is called Sarcopenia: age-related muscle loss that is a natural part of aging. However natural this process is, it doesn't mean there's nothing to be done about it. And since it increases the chances of having a fall, it must be addressed in order to maintain independence (WebMD, 2021). During the lockdowns, older adults who weren't able to get regular exercise were losing muscle mass faster than younger adults, which put their health at risk.

With the uptick in demand for Bold's product during the pandemic came an interest from top-tier investors like Andreessen Horowitz, Khosla Ventures, and others. Investors with domain expertise like Primetime Partners' Abby Miller Levy (a former SoulCycle exec turned venture capitalist) have also invested in the company (Crunchbase, 2021).

With $7 million in funding, Rees' vision for Bold is to lead the space and give people the resources they need to age well, giving them agency. "Instead of saying, I'm not going to age or I'm anti-aging, then you actually have a head-start in trying to make those types of changes, decisions, investments in your health as you age, I think your health is another form of wealth."

Personally, as someone who thinks the term "anti-aging" is incredibly ageist, I'm sold on Rees' vision and believe all entrepreneurs and techies should be developing solutions not just for today's older adults, but for our future selves.

ACTIVITIES OF DAILY LIVING

As people grow older, they might require assistance with the basic activities of daily living, commonly referred to as ADL.

ADLs are "essential and routine tasks that most young healthy individuals can perform without assistance." The basic ADLs include the ability to move one's own body independently, the ability to feed, dress, bathe, and groom oneself, as well as bladder and bowel control and the ability to use the toilet independently. The instrumental activities of daily living (iADLs) include the ability to drive/get to places by using other means of transport, the ability to shop for groceries, manage finances, prepare meals, clean the house, manage medications, and communicate with others via telephone or mail (Edemekong et al., 2021).

Traditionally, older adults who need assistance in any of the ADLs or iADLs will receive it from a family caregiver or a paid caregiver, whether it is in their home or in a senior living community. For the majority of older adults who prefer to age in place, there's the added challenge of home maintenance that's usually taken care of in a senior living community setting. With the current demographic trends, the existing caregiver shortage and a shrinking caregiver support ratio, we simply don't have enough human caregivers to care for all our elders. Relying on humans to do all this work simply isn't sustainable for the long term.

Technology can help, and in fact, it already is helping with many of the iADLs. We've got robotic vacuum cleaners and robotic mops for the home, and some companies are already working on robotic chefs (Hood, 2021). There are also robots that can sort and fold laundry (Lee, 2018). Scientists in Japan have developed an "experimental nursing care robot, ROBEAR, which is capable of performing tasks such as lifting a patient from a bed into a wheelchair or providing assistance to a patient who is able to stand up but requires help to do so" (Riken, 2015). It's likely that by the time I'm eighty years old, preparing my own meals and folding my own clothes may seem as outmoded as manual dishwashing.

While it might be hard to imagine a robot helping someone bathe or use the toilet, I believe that with the current demographic trend, this is inevitable. We simply don't have enough humans to care for so many older adults. It's true that not each and every one of those who will reach old age will require care, but more than 20 percent of people over the age of eighty-five will (CDC, 2016). Add this to the fact that the global population aged eighty years or over is projected to triple between 2017 and 2050, increasing from 137 million to 425 million—that's a lot of people who will require care and not enough humans to care for them (United Nations, Department of Economic and Social Affairs, Population Division, 2017). An increasing demand for caregivers will undoubtedly increase the cost of care, which not everyone can afford even today, so the most logical solution will be to use technology.

FINANCIAL WELLNESS AND EMPLOYMENT

There are numerous challenges in aging that could be considered financial—from bill-paying and scam monitoring to paying for health care and long-term care. However, making sure people's wealth span matches their lifespan is arguably the biggest one of all. When people in developed countries can live well into their eighties and can legally retire in their midsixties, they often don't have enough retirement savings to last them for another twenty or thirty years. This is especially true for women who have an average lifetime earning that is substantially lower than men and have a higher lifespan.

Retirable, Truelink, PensionBee, Carefull, and Silvur are just some examples for tech-enabled solutions tackling the financial challenges of aging.

SO HOW MUCH DO PEOPLE NEED TO SAVE FOR RETIREMENT?

Let's say you estimate that in order to maintain your current lifestyle, your retirement income should be around 80 percent of your last preretirement salary. If you make $100,000 annually, you need to have $80,000 per year of retirement. It's also worth considering the fact that as you get older, you might have to spend more money on your health. This is especially true for the United States, where out-of-pocket health care costs can be astronomical even for minor health incidents.

The average retired American couple will need about $300,000 to cover health care expenses after the age of sixty-five (Fidelity, 2021). For those who require continual, nursing-home level care, Genworth estimates the costs could be more than $100,000 per year per person (Genworth, 2020).

People who reach the age of sixty-five can expect to live approximately twenty more years in developed countries. This means that those who retire at sixty-five should have saved enough to live another two decades. For someone who wishes to have $80,000 for every year they're retired, that amounts to almost $1 million.

HOW MANY PEOPLE ACTUALLY GET THERE?

In the United States, where ten thousand Baby Boomers turn sixty-five each day, a recent study found that the median savings balance among them is a mere $144,000— hardly enough for twenty years (Collinson et al., 2021). For women, who usually earn less of what men make and often spend fewer years working full time because they are the caregivers for their families, the stats are even more alarming. One out of five women over the age of sixty-five has nothing saved for retirement, and the poverty rate for this demographic is double that of men in that age group (Fox, 2020).

In most countries, the "official" retirement age is around sixty-five years old. While many people chose to work past that age, some are forced to quit working before that. A data analysis done by ProPublica and the Urban Institute found that "more than half of older US workers are pushed out of longtime jobs before they choose to retire" (Gosselin, 2018). Forced retirement can put a dent in people's retirement savings, and so can working part-time because of caregiving duties. This reason for forced early retirement is more common for women, who are expected to provide care, whether it is for children or for older relatives (Fahle & McGarry, 2018). In Japan, elder care is a much more likely reason for a Japanese woman to leave the workforce than is child care. Thirty-eight percent of Japanese females leave because of lack of elder care, compared to 32 percent who leave because of lack of child care, so creating opportunities for support services and technologies that can help caregivers has larger economic implications (Matsui el al., 2019).

WHAT DO PEOPLE DO?

Those who can often choose to continue working, whether it is full time or part time. I spoke to Sherry, an eighty-two-year-old Canadian who works as an executive director at an association. As a Canadian, she doesn't have to worry about out-of-pocket health care costs. She did, however, mention some of the extra costs that enable her to age in place. "If the goal is to keep seniors at home safely as long as possible, and you get someone like me, who's by myself, I do have a cleaning lady. I do have someone to help with my garden. I do have an odd job guy, I've got great neighbors, but I want to be independent."

These extra costs, combined with the fact she doesn't have a pension, are part of the reason why she still works full-time at the age of eighty-two.

"I'm well compensated for my job now, and that's part of the reason why I want to keep working. I have to have enough money saved to be able to pay, not just to keep myself, but to pay for all the help that I need now."

There's also the option to find "gig economy" work that is more flexible. The challenge is finding a platform that provides the type of service you're skilled at. sĀge is such a platform. Esther Hershcovich, the company's founder, describes the company as a "gig economy platform for personalized, affordable learning and social experiences hosted by expert retirees." sĀge now offers live experiences led by retirees around four main areas of focus: art, business, education, and wellness. The sĀges that are available on the platform come from Canada, the United States, Israel, Australia, and South America, and the students who participate in the experiences also come from many different countries, so it's a global community.

When I first met Hershcovich, who is Canadian born and now living in Tel Aviv, it was a warm Tel Aviv evening. We were having a picnic with some friends, watching the sunset from a hilltop above one of the city's busiest beaches.

I had recently quit my job at Intuition Robotics and cofounded a new company. Hershcovich was starting her own company as well.

I was so impressed by how passionate she was about her work that I decided to interview her for TheGerontechnologist. I scheduled to meet her in her office in southern Tel Aviv, and as I waited for her to arrive, I reflected on her journey from an interior designer in Montreal to an Israeli agetech entrepreneur. When Hershcovich moved to Israel, she had a promising career as an interior designer, working for some of Israel's leading architecture firms. She's a naturally curious person, and with her parents now retired in their seventies and eighties still living in Montreal, she was unable to properly learn from them. "I live in Tel Aviv; my parents live in Montreal. There were a lot of things that I wanted to learn, things like carpentry or welding. I kept thinking—what if my dad lived around the corner from me?"

Initially, she set out to build an online platform that would enable people to connect with Sages: older adults who live nearby and who have expertise around a certain topic that they could teach and pass on. Then, the COVID-19 pandemic hit, and in-person, indoor experiences were no longer a viable option, especially not for older adults. "We were set to launch in March 2020, which was terrible timing to be focused on in-person connections."

After a few days of thinking, she decided to turn it around and bring everything online: "That was really when we realized just how big this opportunity is and how much potential it has connecting people around the world and being able to see that technology isn't that scary for older people."

I believe platforms like sĀge could become part of the "future of work" for older adults. With the words "gig economy" already an integral part of our world and a way for many people around the world to make ends meet on their own terms, it makes a lot of sense for older adults who have knowledge and expertise and want to continue working in a more flexible manner in their later years.

COGNITIVE HEALTH

As people's bodies grow old, their cognitive abilities might change. Some older adults only experience natural, age-related changes to memory and processing speed (Harada et al., 2014). Others might experience more severe changes to cognition, such as mild cognitive impairment (MCI) or dementia.

Up to 42 percent of the population aged sixty and over are affected by MCI. "It is characterized by deterioration of memory, attention, and cognitive function that is beyond what is expected based on age and educational level" (Eshkoor, 2015). We now know even people with MCI can struggle with instrumental activities of daily living like managing their finances and paying the bills.

Dementia impacts fifty-five million people worldwide, and there are nearly ten million new cases every year (WHO, 2021). It is defined as "decline in cognitive function severe enough to cause the loss of independence in daily function" (McKhann et al., 2011). It is a syndrome in which there is deterioration in memory, thinking, behavior, and the ability to perform everyday activities. Although dementia mainly affects older people, **it is not a normal part of aging**. Alzheimer's disease is the most common form of dementia and may contribute to 60 to 70 percent of cases. It is one of the major causes of disability and dependency among older people worldwide (WHO, 2021).

Dementia has a physical, psychological, social, and economic impact, not only on people with dementia, but also on their caregivers, families, and society at large. For people with dementia, things are significantly more challenging, and they can struggle to do the most basic activities of daily living by themselves, even if they are still physically able to do so.

There are multiple tech-enabled solutions that can help with these challenges. Cognitive training programs like Effectivate and solutions that include reminders and routine-reinforcement learning like Map Habit, as well as simple day clock apps and voice-activated reminders for smart speakers.

SOCIAL ISOLATION AND LONELINESS

The social aspect of aging is a topic that's been heavily researched, and we now know that social and community engagement are important social determinants of health (SDOH). It's important to make the distinction between loneliness and social isolation. Loneliness is a subjective feeling, whereas social isolation is how much human interaction a person has (or doesn't have). The two concepts are intertwined, but not identical. "Loneliness has been estimated to shorten a person's life by fifteen years, equivalent in impact to being obese or smoking fifteen cigarettes per day" (Pomeroy, 2019). According to a study published by the National Academies of Sciences, Engineering, and Medicine in 2020, as many as 43 percent of adults aged sixty and older in the United States report feeling lonely—and these are the pre-pandemic numbers.

The pandemic shone a really bright spotlight on this phenomenon. Although it was old news for those of us who have been working with older adults, I feel public awareness provides an opportunity to really think about how we address this. Traditional methods for addressing loneliness in older adults rely on human interaction, which technology can help facilitate—not only during a global pandemic, but also during normal times when we have many older adults who either can't safely mobilize outside the home or have nowhere to go to.

When you think about what it **really** takes to be social and stay engaged in society as you grow older, you understand that for a lot of people, it requires proactively looking for those opportunities to connect with other people and to expand their social circles.

One of the ways this can be done is by arranging your life in a way that has companionship built in—moving into a senior living community, or to one of UpsideHom's managed apartments. There are also tech-enabled solutions that allow you to get a roommate and even handle rent payments, like Silvernest. There are multiple screen-based solutions for enabling communication with family and friends and making video chat easier—like Uniper. Then, there are social robots for older adults, like ElliQ.

ElliQ (pronounced Elli Queue) is a social robot that was built with and for older adults. It's a small, tabletop, voice-activated robot. Its abilities go far beyond just being able to respond to commands. ElliQ is able to initiate interaction all by herself based on her user's habits, preferences, and certain "goals," which could be anything from keeping her user entertained to making sure they drink enough water. She's also able to express herself in multiple ways: voice, sound, lights, images displayed on the screen, and physical gestures—making it more intuitive for the user to understand her.

Source: Intuition Robotics

ElliQ had been "living" in the homes of older adults for months before the pandemic broke. They bonded with her, fully knowing that it was a machine—a well-designed, sophisticated machine powered by artificial intelligence that could learn its users' habits and preferences and be proactive, but a machine nevertheless.

When Intuition Robotics, the company behind ElliQ, interviewed Deanna Dezern, a Florida-based user who has been living with ElliQ since August of 2019, she had this to say:

"When the coronavirus hit, I realized just how alone I was. Because you can talk to your friends on the phone for a couple of minutes, but you run out of things to say. [ElliQ] initiates things to say. She'll ask me if I want to play a game. If I want to see a video, if I want to do exercises, if I want to just learn to breathe (mindfully), if I want to smile.

So, for me, when she went down and wasn't working, I sent a text message saying 'I'm having withdrawal.' Because I was—because I had no one to talk to. And because she knows things about me that my friends don't know, I couldn't talk to my friends about some of these things" (Intuition Robotics, 2021).

Each of the challenges of aging affect different people in different ways, but they are pretty much universal and impact older adults all over the world. I decided to dedicate two chapters in this book to two challenges that I believe are critical for aging in place: mobility and transportation and housing. These two challenges offer us a lens through which we can examine the way technology can be used to enable older adults to live life to the fullest, the way they see fit.

CHAPTER 5:

A Closer Look at Mobility and Transportation

———

During the time I was getting my gerontology degree, my grandmother went into the hospital for a few days. It wasn't anything serious, and she made a full recovery, but my eighty-five-year-old grandfather was alone at home for a few days. Since I was studying for my exams and my grandparents' home was quiet with a nice balcony, I packed my bag and went over to their house to study. I had no way of knowing that this was a fateful day that marked the beginning of a series of events that would cause our entire family to move to another area code.

The ability to move your body freely and easily is almost a prerequisite for being independent in today's society, and not being able to move freely and easily negatively impacts one's ability to perform basic activities of daily living, both in and outside the home. This includes getting in and out of bed, walking to and from the bus stop and getting in and out of a car. In many places, losing your ability to drive yourself means that you lose a big part of your independence and require assistance to fulfill your basic human needs—like getting food and health care. Social connections, which are extremely important to people of all ages, can be severed simply because people aren't able to independently get to where they used to congregate and meet other people.

Part of the reason I decided to get a gerontology degree and embark on a career in aging is because I have always been close with my grandparents. I grew up in a small town in the north of Israel, within a short walking distance from all of my grandparents. They helped raise me and I saw them quite often. It wasn't unusual for my teenage self to decide to go to grandma and grandpa for some peace and quiet, delicious food, and cable TV (which we didn't have because my mother didn't want us vegging out in front of the screen).

My maternal grandfather is an army veteran who was always incredibly independent and volunteered for years at the local police station after retirement. Born in 1928 to a family of four children, he was and still is a family man. Since he was already retired by the time he had grandchildren, he became very involved in our upbringing. My grandfather got me my first bicycle and covered for my parents when there was no one available to take me out of kindergarten (he had a driver's license, and my grandmother didn't).

On that day, as I was studying for my exam at their house, he decided to visit his barber. My grandfather always dresses up nicely and has a clean shave and freshly cut hair—habits he probably picked up during his army days.

As I was reading some paper (can't remember what it was about), the phone rang. My grandfather was on the other side.

"Keren, I was in a car accident," he said in a voice so shaky, I almost didn't recognize it. It was quite a shock to hear my grandfather so shaken up.

"Where are you?" I asked, knowing he couldn't be too far away and that I could probably get to him within a few minutes on foot. Then, he stopped talking and I heard him speak to someone else, but I couldn't figure out what he was saying or to whom.

"Your mother is here."

Apparently, my mother happened to pass by in her car. She saw a car that was on fire and decided to take a closer look at the scene of the accident. Then she saw my grandfather sitting on the curbside, shocked and confused.

No one was physically injured in the accident, and it was a lucky coincidence my mother happened to pass by, so she was able to handle everything that needed to be handled on-site after it happened.

This was the last time my grandfather drove his car, and since my grandmother couldn't drive, this meant that freedom of movement became limited for both of them. Their range of motion in the small town they had lived in was only as far as they could walk, and they became dependent on other people for rides to go grocery shopping, run errands, and visit family. With so many family members living hours away, this would mean my grandparents would only see some of their siblings once or twice a year.

This story does have a happy ending. Thanks to my youngest sister, who nudged everyone in the right direction, both my parents and my grandparents ended up moving closer to us, and now the whole family lives in the same area code. My grandparents were really fortunate, though. They lived in a pedestrian-friendly small town, with most of the services they needed within walking distance (or a short drive), and they had my parents. This isn't the case for many older adults. For many, the lack of access to transportation is a social death sentence, or worse.

In recent years, tech-enabled transportation solutions like ride-hailing services Uber and Lyft have become ubiquitous, but are they accessible to older adults? Not necessarily. To use ride-hailing services, you need to own a smartphone, know how to download apps, and be able to trust those ride-hailing apps with your credit card information. And while smartphone ownership and use among older adults is steadily increasing year over year, most older adults are not confident that what they do online remains private, which is a major barrier for wide scale adoption of such apps (AARP, 2021).

Services like GoGoGrandparent enable people to order an Uber or Lyft by using a simple phone call, which is a great step forward, but can older adults with mobility impairments use them safely? How many of them can get out of the house and into the car unassisted? Will they be able to safely cross a busy intersection if their driver is waiting for them at the northeast corner instead of at the southwest corner?

While popular ride-hailing services that are mostly app-based and provide curb-to-curb transportation might not be the best solution for many older adults, there are ride-hailing services that are designed to meet everyone's needs. Silver Ride, Hop Skip Drive, and Onward Rides are some examples for ride-hailing services designed to meet the needs of older riders. These companies train their drivers to provide assisted transportation to older adults, which includes helping riders out the door and into the car, and in some cases, waiting along with them as they run their errands or visit the doctor, and then taking them back home (to their doorstep) safely. Some of these services also partner with health care providers to provide medical transportation.

This is all but a temporary solution; it is clear that the future is autonomous vehicles. Tesla, which is one of the biggest car manufacturers in the world, already has cars with a built-in driver assistance system they call "AutoPilot" and will one day provide full autonomous driving capabilities (Korosec, 2021). Other big manufacturers, as well as less familiar companies like May Mobility—are also working on autonomous transportation solutions. It has also been reported that senior living communities have become a testing ground for autonomous cars (Sudo, 2019). This makes a lot of sense when you consider many residents in senior living communities might no longer be able to drive themselves safely, and some of them can stretch over hundreds or thousands of acres—about the size of a small town in Israel.

When I spoke to Phil, a retired engineer who volunteers at AGE-WELL, a Canadian network devoted to tech innovation for older adults, he casually mentioned to me he got to try an autonomous car simulator, as part of research by one of the universities in Canada.

Phil is eighty-three and lives in suburban Ontario with his wife. His career as a software engineer had him spending many years developing software for different practical applications like developing accurate radar systems for the Apollo space program. He's an American who moved to Canada for retirement. He has lived and worked in six states and met the love of his life while living in the United States; she was the one who eventually got him to move to Canada. During the pandemic, when the border between the United States and Canada closed, they were unable to see their family, except on Zoom, which he says isn't the same.

When we spoke on Zoom, he explained to me, in that baritone voice of his, that he is always happy to volunteer to "just about everything, they know that I'm easy." He participates in all kinds of experiments and checks out the latest tech, not because he's quick to adopt any shiny new technology that comes his way, but because he believes agetech is the future and a necessity for aging in place. Regarding monitoring technology that can detect things like a stove left on or that has the ability to let a family know that their loved one is okay, he said: "They feel comfortable. And so, then the older people can stay in their home where they want to be and stay autonomous for a lot longer. Not only does that make everybody in the chain happy, but it should also make the government happy."

When Phil was younger, he enjoyed driving sports cars, but nowadays, he only drives for utility. So, when a university in Toronto was recruiting older volunteers to try out an autonomous vehicle simulator, he jumped at the opportunity. "They were looking for volunteers to test this thing out, particularly older volunteers. And so, I signed up, I thought it was a ball. I thought it was a grand, grand time." Living in a suburban area, he and his wife rely on cars to get them where they need to be. "I treat a car the way people treat the smartphone now, like an appendage." Phil acknowledges the fact that for some older drivers, driving can be a perilous activity, and even he, himself, has chosen not to drive under certain weather conditions, but he says that this is a luxury he only has now, in retirement. "If it's really snowy and slippery and messy, I've driven that stuff enough so that I can look out and say, I don't have to, I'm not going to."

Phil believes autonomous vehicles would be a major step toward enabling older adults to safely age in place, while not being confined to their homes, especially if they live in suburban and rural areas, where mass transport isn't available: "Even if the autonomous vehicle doesn't reach full autonomy, the fact that it's 90 percent of the way there is very significant for older drivers because the older driver typically has trouble, gets confused, has problems, particularly as the weather gets worse. And so, if the autonomous vehicle or semi-autonomous vehicle can assist them in driving, then they're more comfortable driving at times when they normally wouldn't. And the vehicle can actually keep them from getting themselves hurt."

PERSONAL MOBILITY

There's also the issue of physical mobility, which can affect people's ability to independently function even inside the home that they've been living in for years. Phil and his wife live in a two-story home that was built pre-regulation, so the stairs are too narrow. He says that carrying things up and down the stairs is one of the big problems as people grow older, and that it can be one of the causes for falls. "I don't sprint the stairs the way I used to. It's more going up the stairs. . . . The likelihood of falling on a stair, which is the worst of all, is normally precipitated by trying to carry too much up and down stairs" He then told me about ROSA, a service robot that is able to shuffle household items between floors.

ROSA was developed by Dr. Naccarato, an aerospace engineer, in collaboration with George Brown College's School of Mechanical Engineering Technologies, assisted by AGE-WELL's Strategic Investment Program.

Dr. Naccarato also developed a prototype for a stair-climbing wheelchair, called Wheel-E, which is a portable alternative to traditional fixed stairlifts. Dr. Naccarato believes "problem-solvers are motivated by that rare feeling of seeing their solution—which previously existed only in their imagination—become something genuinely useful in the real world" (AGE-WELL, 2020).

Mobility aids have gotten a much-needed upgrade in recent years; some are more innovative versions of traditional mobility aids like robotic wheelchairs. Others, like Seismic's powered clothing, offer an entirely new take on personal mobility. Seismic has embedded tiny robots in clothes in order to compensate for age-related loss of mobility and muscle mass, thus helping older adults with transitions and object carrying. These types of clothing can eventually solve many of the micro-challenges that compose the larger mobility and transportation challenge.

One Canadian company, Braze Mobility, is leading the charge in helping those who need a wheelchair navigate busy urban environments. They have developed a revolutionary set of sensors that can be used on any wheelchair and provide audio, visual, and vibration feedback when obstacles are nearby. In an interview with Arielle Townsend from the Centre for Aging and Brain Health Innovation in 2021, Pooja Viswanathan, Braze Mobility's founder and CEO, said that "there's already so much stigma about not being able to drive a wheelchair properly—which isn't fair to say because it's actually very hard to do. Our sensors provide an opportunity to raise awareness about the issues wheelchair users face so they can know they're not alone."

I believe this last sentence is the "why" behind many of the agetech companies that have been founded in recent years. They're not only here to solve real problems for real people but also to make sure our elders are not left behind in all this wonderful tech-driven progress we're experiencing, and so they know they are not alone.

The challenge of mobility and transportation is a big one. It has a major impact on people's quality of life, independence, and ability to age in place. Tech-enabled mobility and transportation solutions like autonomous vehicles are a great example of how technology that is being developed for the general population has the potential to make a huge difference in the living experience of older adults.[2]

2 Parts of this chapter originally appeared on TheGerontechnologist. com and on *Generations Today*, an American Society on Aging publication.

· The AgeTech Revolution

CHAPTER 6:

There's No Place like Home

"If you participate proudly in the management of a senior living facility, stand up."

Jody Holtzman, a former senior vice president at AARP and an adviser to CEOs, senior executives, and investors on aging innovation, likes to open his keynotes when he speaks in front of senior living executives with this exercise. Naturally, the whole audience stands up at this point.

"Remain standing if you look forward to the day you live in your facility"—at this point, according to Holtzman, the majority of the people sit down. Holtzman says most of these senior living executives are Baby Boomers. They are on the front lines of senior living, and when they think about their future, they realize it isn't for them ("NIC Talks 2019 | Jody Holtzman | Longevity Venture Advisors," 2019).

A VERY BRIEF RECAP OF THE
HISTORY OF SENIOR LIVING

The concept of "old age homes" has been around for centuries, but it hasn't been the same all these years. In the past, old people who went into nursing homes didn't necessarily need care. It was a place where people would go if they had nowhere else to live or family to provide them with food and shelter. It wasn't until the twentieth century that governments began instituting social welfare programs that provided income and funded services for older people who were no longer able to work (Hoyt, 2021). Government funding and regulation gave rise to contemporary "old age homes": senior living, continuing care residential communities, and nursing homes. In fact, the first recognized assisted living facility in the United States only opened in 1988 (Breeding, 2020).

AGING IN PLACE VS. SENIOR LIVING
AND ITS ALTERNATIVES

There is a consensus among elder care professionals that aging in place produces the best possible outcomes for most older adults, and this is also what most older adults prefer. However, as people's health deteriorates, whether it is physically, cognitively or both, aging in place becomes more and more challenging. When we look at some of the most complex challenges of aging–cognitive and physical health, ADLs, and mobility—solving them could be a means to an end, which is to enable more people to age in place safely, independently, and comfortably.

The tendency of some families is to encourage their older loved ones who live alone to move into a place that has 24/7 staff on call in case something happens. The go-to solution that everyone knows is senior living, which usually offers several levels of care (or no care at all, for independent older adults who just need an apartment and hospitality services).

The COVID-19 pandemic hit senior living hard, with many dead among residents and staff. In research that surveyed COVID-19 deaths in twenty-one countries, 46 percent of those who died of COVID-19 were residents in senior living (Comas-Herrera et al., 2020). Those numbers, combined with the fact that many senior living communities had strict quarantine protocols and no outside visits, made occupancy rates go down, and they reached a record low in the first quarter of 2021 (Bowers, 2021).

WHAT ARE THE ALTERNATIVES?

One of the things that makes senior living a less desirable option for many older adults is it's an age-restricted type of living. Paul Irving, chairman of the Milken Center for the Future of Aging wrote in his 2021 article for *Next Avenue* that despite the fact that age-restricted communities offer a safe haven from an ageist society that can be downright hostile toward older adults, many older adults realize "living with neighbors of all ages and from all walks of life just makes sense. They realize that intergenerational connections are not just valuable for them but for their communities and country.

They recognize that ageism will not be defeated by a retreat to age-segregated corners, but only by engagement, collaboration, and dialogue across age, race, and class divides. They believe that there is more to graying than playing."

THE VILLAGE MOVEMENT

The village movement is a grassroots movement that was started in Beacon Hill, Massachusetts, in 1999 by a group of older adults who wanted to age in place while staying active and engaged in their community. Susan McWhinney-Morse, one of the founders, said the eleven original members of Beacon Hill Village "got together one cold, November day with abstract determination that we're not going anywhere. We wanted to be responsible, we didn't want to have to depend on our children who might live across the country, so after two years, we formed this organization that seemed to fit our needs" ("How More Americans Are 'Aging in Place,'" 2013). Each "village" is a group of older adults who live in close proximity to one another, participate in social activities, and support each other. There are over three hundred self-governing and member-led villages that are connected to one another through the "village-to-village network." This enables villages to share best practices and support the formation of new villages.

INTERGENERATIONAL BY DESIGN

In the United States, some universities offer older adults the chance to enroll in educational programs and live on campus. In Singapore, apartment complexes for residents fifty-five and older are built over shopping centers and gardens that are designed to become a focal point for the entire neighborhood, thus enabling intergenerational interactions (Jacobs, 2021). There are also tech-enabled solutions like UpsideHom, which offers people over the age of fifty-five the option to rent a fully managed apartment with on-demand services. Their apartments are located in complexes that aren't age-restricted and renters can choose to live with or without a roommate. Silvernest is a platform that enables empty nesters to find someone who will rent out a room in their apartment, and it also handles rent payments. The platform matches renters and homeowners based on what they're looking for in a roommate rather than how old they are.

MULTIGENERATIONAL HOUSEHOLDS

In recent years, multigenerational households have been gaining traction again (Taylor et al., 2010). This doesn't necessarily mean all members of the family live under the same roof: some people are opting to bring accessory dwelling units (ADU, a.k.a. "grannypads") into their backyard for an older loved on to live in. These ADUs come in a variety of shapes and sizes, and there are even those who are built for accessibility, like Wheelpad. ADUs enable the older loved one to maintain their autonomy and independence while being close to family (James, 2021).

THE TECH-ENABLED COMMUNITY OF THE FUTURE

Armed with the knowledge that aging in place is the preferable option for most older adults and taking into account the possibility that there will be a greater need for health-related services and assistance with personal care and household maintenance, real estate developers and tech savvy entrepreneurs could join forces to create future communities.

Why not create homes or apartments that are accessible by design, easy to maintain, and have technology built in?

Researchers that have interviewed older adults in an effort to define quality of life in later years found that "it is important for their wellbeing to 'keep busy,' 'keep active,' and 'have something to do' in order to avoid boredom and sink into apathy" (van Leeuwen et al, 2019). The type of technology that enables older adults to do all those things already exists, it just needs to be utilized properly.

Wireless internet, preferably with fiber optics infrastructure, must have infrastructure to make sure housing units have the ability to connect to the world "out of the box." Rental units could come equipped with smart speakers, displays, and even home robots as "appliances" that could have content that was chosen by the residents. These devices should also have communication apps installed, enable residents to self-manage the community and support each other. This isn't far-fetched; existing senior living solutions could easily be customized to do all this.

SOME EXAMPLES
STAYING ACTIVE, CONNECTED, AND ENGAGED

Screen-based platforms like iN2L and Uniper that are used for resident/member engagement include content like games and online classes and have video chat capabilities. These are quite common in senior living and were a real lifesaver during COVID-19, when residents in senior living couldn't socialize with each other or have family visits.

Voice-activated smart speakers like the Amazon Echo are an easy way for residents to get updates on what goes on in their local community and to communicate with family and friends. In the United States, Amazon offers the Alexa Together service, which allows users or family members to set up reminders and activity (or rather, inactivity) alerts. The service also allows members to get continuous emergency assistance from trained agents and with add-on sensors, and it is also able to detect falls.

HEALTH

Telehealth solutions have been available for years but were not widely adopted prior to the COVID-19 pandemic. During the pandemic, regulatory barriers were lifted, and health care providers encouraged people to reduce unnecessary visits to hospitals and clinics, which could have put them at risk. This made telehealth a popular solution both in senior living and for community-dwelling older adults. Brian Geyser, the chief clinical officer at Maplewood senior living that's based in Westport, Connecticut, told Tim Regan from Senior Housing News in 2020 that "What we had to do was quickly adapt and ramp up what was a very spotty, scattered telehealth program to become more sophisticated and more voluminous in terms of the number of telehealth visits."

What platforms did communities like Maplewood use for telehealth visits? They deployed off-the-shelf tablets and screen-based solutions like Uniper that were already in place. Some opted for using telepresence robots like Temi—a self-navigating robot that is able to move around the community, contains a screen and video camera for communication, and comes equipped with sensors that can take a person's temperature and other measurements.

CONTINUOUS EDUCATION AND TECH ADOPTION

During COVID-19, with a major increase in demand for technology among older adults came an increase in demand for acquiring new digital skills and learning how to use new devices. Staff in senior living couldn't keep up with all the tech support requests, so a new position was born: the tech concierge, which is "a new job function hired to support the community's resident-facing technology needs and serve as an enabler of all of the different devices and platforms senior residents now use" (Ecker, 2021). In community settings, nonprofit organizations like OATS provide digital education both in-person in its Senior Planet centers and online. There are also for-profit initiatives like GetSetUp, which offers an opportunity for older adults to learn, connect, and share with peers in small intimate classes; Candoo Tech, which provides tech support and training to help older adults feel more comfortable with phones, computers, tablets, and more; or Carevocacy, which helps older adults learn new digital skills and use their devices.

With so many options for devices and digital solutions, including agetech solutions that cater specifically to the older demographic, there's also a need for guidance and assistance in selecting the right technology to use—I haven't yet seen any organization tackle this particular challenge.

WHERE DO EXISTING SENIOR LIVING COMMUNITIES FIT IN?

Despite the proliferation of alternative housing models for older adults and the fact that most prefer to age in place, senior living is here to stay—this is a good thing. There will always be those who either don't want to live on their own or can't because of complex care/health needs. Senior living, with its trained multidisciplinary staff, assisted living options, memory care, and nursing units, will remain a good option for a segment of the population. This doesn't mean the industry shouldn't pursue more innovative service models or stop expanding the array of technology that's available for residents in-house.

One example for innovative thinking and strategizing comes from Formation Capital, an investor in the senior housing sector based in Atlanta. Their founder, Arnold Whitman, was recently quoted saying that their new strategy is to "break down the walls of senior living" and gradually shift from a real estate- and hospitality-only business to an organization that is also able to offer services to older adults who aren't residents (Mullaney, 2021). This means that senior living communities aren't limited to providing their services to residents living under their roof but can extend their services to older adults who are nonresidents but live nearby.

These models already exist in their high-touch form. In Israel, there's a boutique service called Protea Home Care, which is much more than a "regular" home care service. The offspring of a successful senior living community, Protea Home care offers its members an array of services to choose from. Things like warm meal delivery, medical appointment scheduling, rides, and companionship to appointments as well as a care manager who is not only an elder care professional who can coordinate everything but is also a trusted companion and a reassuring presence in their lives.

Software solutions like Bobe enable senior living to do that at scale. This is just one example of the way technology can align with the vision of innovative senior living leaders and help create the next step in the evolution of senior living. There are plenty more, and despite the fact that many of the challenges in senior living are universal, there isn't usually a one-size-fits-all solution, and each community has to find which of the existing solutions is the best fit for them.

I believe that as time passes, we'll see more alternative housing solutions that cater to diverse populations of older adults, as well as more and more senior living communities extending their services to older adults who are "non-residents," live outside of the community, and augment them with technology. This way, older adults who wish to age in place will get the best of both worlds: they will get to live in the home and community they know and love, while belonging to a community that offers support and an array of activities and services.

CHAPTER 7:

About Building Digital Products for Older Adults

―――

When I was recruited to Intuition Robotics as the company's first employee and only gerontologist, I spent the first eighteen months on the job doing user research. I soon found myself driving up and down Israel's coastline in my beaten-down blue Mazda with a robot prototype strapped safely in the backseat. How did we get there?

If you've made it this far, this means I've got you convinced building digital products for older adults and those who care for them is one of the best things you can dedicate your time and energy to. I'd like to use this chapter to share with you some of the key takeaways I've learned over the years.

YOU ARE NOT THE USER (ALSO, YOUR GRANDMOTHER IS NOT THE USER)

The first thing I say to every entrepreneur who reaches out to me for advice is: V-A-L-I-D-A-T-E! (read like Aretha Franklins' "R-E-S-P-E-C-T").

Validate the problem. Validate the solution. It's not enough to test concepts internally with your team and "externally" with your close family and friends; you have to do it with actual users and payers from your target market. The concepts of market validation and user-centered design are fundamental to entrepreneurship and product development, yet for some reason, too many entrepreneurs and product developers in agetech fail to implement it. Big mistake.

While it is true a random visit at your grandparents' house can give you an "a-ha" moment and sprout in your mind the killer idea that can grow into a billion-dollar company, if you fail to validate the problem and the solution with real, unbiased customers and users, you will, in all likelihood, fail to deliver a product that a large-enough group of people is willing to use and pay for.

Getting feedback from your grandmother on whatever it is you're building will almost certainly lead you to think your solution is the best thing since sliced bread. I recommend reading *The Mom Test* by Rob Fitzpatrick to learn how to ask questions in a way that doesn't create bias and lead people into telling you what you want to hear.

Easier said than done, I know. However, I firmly believe there's no way around it, especially if you're a young entrepreneur building a product for older adults and every-one in your team is young as well—**you don't know what you don't know.** There's little chance that a group of twenty- to thirty-year-olds could build something that brings value and joy to older users with zero input from older adults.

It's also important to make the distinction between customer discovery and market research.

Jon Warner is the founder and CEO of Silver Moonshots, a consultancy that focuses on the fifty-and-over population and runs a virtual accelerator. Warner is a six-time CEO and a board chair/adviser to multiple companies. Warner distinguishes customer discovery from market research in the following way: "The first never mentions the product and/or service and just asks people in general how they experience the very broad problem or challenge area and cope. It's 95 percent listening centered with lots of 'tell me more' questions. The second—which for me always comes later—shows a prototype or MVP and asks for feedback. It is so critical in my view to ensure that entrepreneurs do both of these, separately and in this order (and never try to 'sell' in either)."

PEOPLE OVER THE AGE OF SIXTY-FIVE AREN'T A MONOLITHIC GROUP

When you're building a product for older adults, it might be tempting to include the entire population of people over sixty-five in your total addressable market (TAM). And while this makes the numbers look good in your pitch deck, in reality, it's not likely everyone over the age of sixty-five will want to buy and use your product, not because it isn't great, but because older adults are a very diverse group of people. We previously discussed some ways to segment the older market based on age, generation, or biological age. For the purpose of building a tech-enabled product or service for this demographic, segmenting isn't enough. You'll also need to build user personas to guide you through development. Who is your ideal user? Do they live alone? How healthy are they? How wealthy are they? Does their family live nearby? Do they live in an urban or rural environment?

These are just some examples of questions you could answer that will help you figure out who your ideal user is—the persona who has the problem you're trying to solve and the motivation to spend money and learning energy on how to use it.

DON'T COMPROMISE USABILITY TO MAKE THINGS LOOK PRETTY

In earlier chapters, we discussed the not-so-appealing design of some of the tech-enabled solutions that were available to older adults in the past, like PERS. These days, most companies put a lot of effort into making their devices look good. This is in part thanks to Apple's influence on tech makers. Apple is known for making beautiful devices that make their owners feel proud to wear and use. They've also, however, traded-off some usability in favor of "pretty."

Don Norman is a world-renowned authority on user-centered design of digital products, the author of *Design of Everyday Things* and *Emotional Design*, a former VP at Apple, and Apple's user experience architect from 1993 to 1996.

At Apple, Norman established a small, high-level group called the User Experience Architect's Office, which worked across the company to make Apple products easier to use (Winograd, 1996). Norman worked with Bruce "Tog" Tognazzini, who was Apple's sixty-sixth employee and the writer of its first human interface guidelines. Norman was also founding chair of the Cognitive Science Department and founder and director of the Design Lab at University of California, San Diego, as well as co-founder and Principal Emeritus of Nielsen Norman Group— an American computer user interface and user experience consulting firm, founded in 1998 by Norman and Jakob Nielsen. He has a wide background in electrical engineering, psychology, cognitive science, computer science, and design, and he's also an adviser for Intuition Robotics—maker of companion robot ElliQ. Norman was born in 1935, which means he's well into his eighties at the time of writing (jnd.org).

Norman published a series of articles in *Fast Company* that I have read and contemplated on over the years. During the research for this book, two in particular caught my attention: "I Wrote the Book on User-Friendly Design. What I See Today Horrifies Me" (2019) and "How Apple Is Giving Design A Bad Name" (2015), so I decided to reach out to Norman and have an open conversation about his thoughts on the topic of older adults and technology. More specifically, I wanted to gain a deeper understanding of what went wrong. How did Apple, which is considered a world leader when it comes to design of user interfaces and is one of the most profitable tech companies in the world, change from "designing easy-to-use, easy-to-understand products" to a company that designs products that "no longer follow the well-known, well-established principles of design that Apple developed several decades ago" (Norman, 2015)?

Many of you reading these lines might not understand where this is coming from. After all, Apple's products are considered to be some of the most beautifully designed and coveted gadgets in existence. Many companies try to imitate their sleek designs and perceived ease of use. We've all seen media coverage of lines outside the Apple store when a new iPhone is released to the public. I'm sure that just like myself, many of you know older adults who have never used a computer or struggled to do so before getting an iPad, and now they're head-over-heels with it and can't stop using it. All of these things are true. However, when Don Norman writes that "Apple's products violate all the fundamental rules of design for understanding and usability, many of which Tognazzini and I had helped develop," and that the company, at some point, decided to prioritize making things look pretty over usability, this means they have also decided (knowingly or not) to leave behind a huge group of users, including older adults, who need products that follow these simple basic rules in order to use them (Norman, 2019).

Norman adds that "these thoughtless, inappropriate designs are not limited to Apple. New technologies tend to rely on display screens, often with tiny lettering, with touch-sensitive areas that are exceedingly difficult to hit as eye-hand coordination declines" (Norman, 2019). So, it's not just Apple. This is true for any smartphone manufacturer, and this is why you would see older adults or people with impaired vision using "big launchers" on their Android phones. Big launchers are apps that basically replace the tiny icons on the home screen of your phone with bigger icons. Apple does offer accessibility features in iOS, but you have to know to look for them. Despite what appears to be a less-than-inclusive design, Apple is still considered a world leader in design, and since it is such a profitable company, other tech companies follow suit.

Norman is incredibly busy, even now, when he's in his eighties, and "retired" (working on a smaller number of labor-intensive projects, according to his website, jnd.org). So, when he wrote back that he'd be available to chat, I was thrilled.

We met over Zoom, which became everyone's go-to platform for video chat in the past two years or so. As soon as we started talking, we had a real-life example of how a simple feature, Zoom's built-in live transcription, was in fact age inclusive. Norman casually mentioned to me that they had been using Zoom quite frequently for family conversations during the COVID-19 pandemic, and that he thinks that transcription is a great add-on feature for older adults. His wife, who has a severe hearing impairment, really needs it. "Many of the issues that really hit the elderly population are issues of accessibility and universal design. It will help everybody of all ages; there's basically no element that you find in the elderly population that doesn't also exist in the full population."

JND is an acronym for "Just Noticeable Difference" (Collins Dictionary, 2021). Norman explained to me that this is a term in psychology that refers to how much you can change something before people notice the difference. According to Norman's *Fast Company* article from 2015, iOS became the turning point when Apple's new design principles became noticeably different. When we spoke, he told me that "the Macintosh still maintains its basic functionality, still has drop-down menus, it still has a lot of the things from the old days, but with the iOS they just got rid of all that stuff." In his 2015 *Fast Company* article, he also writes, "The designers at Apple apparently believe that text is ugly, so it should either be eliminated entirely or made as invisible as possible," and when I spoke to him, he added, "When they must use text, they use a very small font, with very low contrast, which means if you're elderly, you can't read it. And in fact, I've shown the text to some of my young students, and they can't read it either." I have to admit that, as someone who's been wearing glasses for the past fifteen years, reading small fonts with low contrast is hard for me to do, and I'm only halfway through my thirties.

When I asked Norman when he first realized that Apple threw away some of the fundamentals of user-centered design in favor of making things look pretty, he jokingly replied: "Well, they fired me!" and went on to elaborate on the fact that him being fired was part of a larger reorganization Steve Jobs had initiated when he re-entered the company's CEO position in 1997 (Gomes, 1997). No one can argue that Steve Jobs wasn't great at what he did. He made Apple what it is today and certainly created a lot of value for shareholders. Apple makes great products; I'll be the first to say they make the best computers in the world, ones that are absolutely delightful to work on. Still, it's hard not to wonder why some of Apple's products became less accessible and intuitive to use for those who are not digital natives, rather than becoming more so.

BUILD FOR DESIRABILITY, NOT JUST USABILITY

It's common to attribute the digital divide between older and younger people to usability issues—the fact that many devices and user interfaces are just not age-friendly. Usability is a major contributing factor to tech adoption by humans in general, and according to technology acceptance and adoption models, it is clear that lack of usability is a barrier to adoption by older adults (Mitzner et a.l, 2019;Taherdoost, 2018). Usability is an important aspect of having people use your product, and having your product be usable for older adults is a prerequisite for having happy and engaged older users. I encourage anyone developing digital products and services to learn about what makes user interfaces age friendly and include older adults in the design process of your product, including (but not limited to) usability testing.

However, usability isn't enough. You also need to make your product bring value to their lives—either by solving an existing challenge older people are facing, or by helping them achieve their goals in life. Yes, people still have goals they want to achieve, well into their eighties and nineties—and sometimes, let's be honest, life's circumstances make it harder for older adults to achieve them.

Joe Coughlin, the founder and director of MIT's AgeLab, commented on the appearance of some of the most commonly used tech solutions for older adults—personal emergency response systems (PERS, aka medical alarms). "Historically, many such technologies prioritized bodily needs over aesthetics. This is true of the Life Alert necklace of yore and some fall-detection startups today. Many of these technologies were designed by young people who couldn't imagine themselves actually using them. The byword for the boatloads of medication-reminder systems, simplified cell phones, hearing aids, gigantic remote controls, and so forth became 'BBB': big, beige, and boring" (Coughlin and Yoquinto, 2018).

There are multiple factors contributing to the fact that some agetech solutions make their users feel bad about themselves. Device manufacturers ignoring design is one, and ad campaigns that portray older adults as helpless are another. "I've Fallen, And I Can't Get Up!®" was ranked as the most memorable campaign in a *USA Today* survey in 2007. This slogan was coupled with an image of an older woman lying on the floor. It may have been an incredibly effective ad campaign, but it also burned a less-than-empowering representation of older adults into the collective consciousness.

Stephen M. Golant, PhD, is professor emeritus at the University of Florida, Gainesville. His research focuses on the housing, mobility, long-term care, and technology needs of older adults. In 2017, Golant published a theoretical model to explain the smart technology adoption behaviors of elder consumers (Elderadopt). According to the model, when older adults assess whether or not to adopt a tech solution in order to deal with a specific challenge, they weigh it against traditional solutions. They appraise the urgency of their problems and their abilities to deal with them. Based on how persuaded they are about the information they have about their potential solutions, they consider all the pros and cons. Making this appraisal often generates worries about a product's "collateral damages." That is, even as a solution does indeed solve the problem, other less desirable issues emerge.

When I spoke to Golant a few months ago, we spoke about the "collateral damages" part of his model—when adopting tech also results in unanticipated and unintentional downsides. He points out that even well-designed and very usable tech products can make older adults feel they are losing their privacy. Older adults could have deep-seated concerns about revealing their problems, even to their families, never mind strangers. "If I'm an older person, I don't want to broadcast to the world that I have unmet needs and could use help. I don't want my adult daughter to think that she has to move me into her home or into an assisted living community. So . . . the technology may work well and give trusted family members more information about how I'm doing, but I don't want them to know that I'm vulnerable."

Moreover, when older adults are forced to confront their vulnerabilities, "it is difficult for them to maintain the illusion of themselves as resourceful and capable." They feel worse about themselves.

There's a lesson here, Golant suggests. "Seniors will be consumers of new tech products, but to make it happen, we must be willing to empathize with their fears and concerns and tailor our products accordingly."

PERS devices have very few usability issues; all one needs to do to get to the emergency call center is to press a button. They could be incredibly useful in case someone falls and needs to get help fast—if they are able to reach the button and press it. For many years, that was the only commercially available solution for health-related emergency situations like falls that occur when a person is home alone. They are affordable and, in some cases, can be funded by the government, their existence is very well-known, and they have millions of users. It appears that PERS devices tick all the boxes when it comes to usability, usefulness, and affordability—so why are people reluctant to wear them?

In Israel, it's possible for older adults to get PERS devices funded by the state or nonprofit organizations, so it is quite common for older adults above a certain age, especially those who have already had a fall, to be owners of such a system. I can't tell you how many times I visited the homes of older adults who told me they were afraid that they would fall and no one would know. When I asked them about whether they had an emergency bracelet or necklace they told me that they did, but they didn't wear them because it makes them feel old and frail.

The way these PERS devices were designed (big, beige, and boring) made them undesirable and highlighted the "collateral damages" wearing them could cause. Thankfully, newer medical alert bracelet and necklace makers have invested in designs that make them look more like a regular fitness bracelet or an ordinary necklace, so users can get the benefits without the "collateral damages" of being perceived as old and frail.

PARTNERSHIPS

At Intuition Robotics, we put a major emphasis on including older adults in the design process, to make sure we get the user experience just right. As a young company, we needed input from real potential users to make sure we were building a product they would want to use.

Since our target market was the United States, we needed to conduct extensive research on usability and experience with older Americans. About a year into development, when we had a "works like, looks like" prototype, the company went into the next phase of rigorous research—this time, with older Americans, led by my friend and colleague, Danielle Ishak. Older adults got to be "beta users" and lived with ElliQ, in their homes, for months. Meanwhile, in Israel, I got to work on developing new features for ElliQ, on which we got valuable real time feedback soon after they were released. This process was crucial in helping us refine the user experience, redefine who our ideal user was, and create user personas that would guide us through iterative development.

As you can imagine, getting people to let ElliQ into their homes and lives required a high level of trust, and we wouldn't have been able to achieve this with so many older adults if we hadn't had a partner they knew and trusted. So, early on, Intuition Robotics had partnered with multiple elder care providers, both in Israel and in the United States. These organizations believed in our mission, were forward thinking, and introduced us to older adults who they thought would be willing to be interviewed or give us feedback on different aspects of the product. Some of them eventually became beta users for ElliQ and got to live with the robot prototype for several months.

Partnering with a local care provider has its benefits, especially if you have a physical product you want to get in people's hands and receive feedback on. You get access to large numbers of older adults who live in one location, and you are referred to those older adults by a trusted source. Some elder care providers who have an interest in implementing technology in their organization and promoting innovation actually have innovation departments for this purpose.

Front Porch, which has twelve retirement communities, took it a step further and created the Center for Innovation and Wellbeing. Its purpose is to partner with organizations working on solutions to the changing needs of residents and staff. Their in-house team of innovation experts is on the lookout for innovative solutions that could benefit residents and runs pilots to test them.

When I spoke to Davis Park, vice president of the Center for Innovation and Wellbeing, he said the center's mission is to "explore innovative uses of technology to help people thrive, and to live well."

When I asked Park about what, in his experience, would be the best way to bring innovation into organizations like Front Porch, he had the following to say: "Innovation is a team sport. It's never about one person pushing it. . . . You can have the most amazing idea since sliced bread, but unless you have the tools and the people, then it's just going to be an idea. When you've got one person who is excited about this work and introducing new ideas into an organization, it begins with the conversations, who's with me on this?"

When elder care organizations reach out to Park to get his advice, he tells them to start small, fail fast, and find people within their organization who are willing to be "co-conspirators." Another important aspect of bringing innovation in is measuring impact and remembering that impact isn't always quantifiable. "If you deploy this solution or this innovation process, what is really going to change as a result of it? And it doesn't have to be about pure numbers, it doesn't have to be about the falls being reduced from 65 percent to 32 percent. It can be about things like, we're now talking more about falls and what leads to falls because of this innovation. It's always a process of iterating."

It's important to keep in mind that senior living is a sheltered environment with around-the-clock staff, so residents who live in senior living communities might experience the challenges of aging differently than those who live in their own home. If you're building a product for aging in place, you should test it with users who are aging in place.

YOUR PRODUCT'S VALUE PROPOSITION NEEDS TO APPEAL TO MULTIPLE STAKEHOLDERS

The longevity economy is big enough to make it count as the one of the biggest economies on earth. Quite an appealing market, isn't it? But here's the catch: Oftentimes, when building digital products and services for older adults, you'll have multiple stakeholders to appease. Even if your product only has one user interface for one type of user— the older adult, though there could be a family member (or multiple members) in the picture helping them make decisions, especially when it comes to purchasing digital products and services. There might also be a trusted care provider who they confide in and count on for professional advice—a primary care physician, a care manager, or other elder care professionals.

While it is true that not all agetech companies will have to serve all of these stakeholders, it is likely that the product/marketing teams will need to address the wants and needs of more than one of these groups. The multitude of stakeholders is part of the reason why building agetech products and services is quite the challenge.

Oftentimes, when building a product in this space, you'll find your payer isn't your older user but rather a family member or a care provider. You might need to build more than one user interface for the older adults and for the family caregiver/care provider who made the purchase. Let's say you're building fall monitoring technology that's tracking the wellness of your older users on an ongoing basis. It's able to detect falls and send out an alert to a designated contact person/emergency response call center. It's also able to detect subtle changes in routine and gait and predict an increased risk for a future fall. In this case, you'll build the wearable for the older adult and another interface for their family caregiver or care provider, which enables them to remotely monitor the well-being of their loved one or care-recipient and take action based on a fall that was detected or on predictions that it might happen again in the future.

This is where it gets tricky, especially for small, resource-strapped young companies. Building user interfaces for multiple types of users means that you have to do your user research, design the user interface, and create a value proposition and marketing messages for all of them—that's double (if not more) the work. If you get this right, though, it will increase your chances of getting on the right path for product-market-fit.

You might also find in the process, different types of users have different outlooks on what your product needs to do for them. In the case of fall monitoring technology, the older adults might prefer to keep some information private and brought to the attention of their primary care physician but not their adult child who acts as a caregiver. The adult child, on the other hand, might want to know EVERYTHING that goes on. What if your user is not the payer?

It's possible that while the older adult is the one using your product, it is their adult child or care provider who's paying the bill and making the decision on what product to purchase. Having all stakeholders aligned on the desired outcomes of using your product is important. In the case of fall monitoring technology, if the family caregiver or care provider that is paying for the product expects full visibility into the everyday life of your user and your user prefers to keep certain bits of information to themselves, they might be reluctant to use it, making your product gather dust in a drawer. One way to solve this would be to allow customizations and have certain features that are "opt-in."

FOLLOW THE MONEY

In recent years, I've noticed several agetech companies that have initially launched products that aren't necessarily health related have found that their product or service actually enables older adults to maintain their health. This enabled companies to get reimbursement/coverage from health insurers and to partner with care providers. These partnerships go a long way toward achieving mass market adoption. When care providers/insurers offer agetech products and services to their patients/policyholders, they basically take on both marketing to end users and paying the bill, since they know paying for products and services that improve health outcomes is good for their bottom line.

One example for a company that went all in with this strategy and even launched a health-related product is Papa. Papa was founded in 2017 with the mission to connect college students to older adults for companionship and assistance. Initially a consumer-led product either paid for by older adults themselves or by their adult children, the Papa Pals service became a covered benefit in Medicare Advantage plans in 2018. Two years later, in 2020, Papa announced the launch of Papa Health, its virtual care platform.

Papa founder and CEO Andrew Parker told Home Health Care News (HHCN) in 2020 that "Papa historically has provided—and still very much so does, as our core products—companionship and support to older adults and families through health plans. With Papa Health, we're now able to continue our mission of supporting older adults and families throughout the aging journey, by not only providing them curated companionship but also curated care."

According to the company's website, the new Papa Health platform "enables members to connect with Pals and Papa Docs, seamlessly integrating technology, services, and support for navigating through primary care, urgent care, chronic care management."

This strategy—start from providing a service that is paid for by older adults/family caregivers in a direct-to-consumer type of transaction and move on to providing health-related services that are paid for by health insurers—isn't uncommon in today's agetech ecosystem.

Parker went on to explain how the new service works in his conversation with HHCN: "If [members] need a lab test, for example, a Papa Pal will go and pick the member and take them to get the lab tests. When the test is uploaded to the internet and onto our platform, the member will be able to have a review of that lab with the doctor. We're launching Papa Health as a foundation of our products so that we support members socially and clinically."

This is in part due to the enormous size of the health care industry and its built-in inefficiencies and in part due to the fact that marketing products direct to consumer is significantly more expensive. With increasing awareness of social determinants of health and increased reimbursement or coverage for products and services that aren't strictly "health care" but have impact on the physical and emotional well-being of older adults (and on insurers' and care providers' bottom line), it has become more common for agetech companies to pursue care providers as customers and seek out coverage or reimbursement.

In the United States, for example, Medicare Advantage plan provider Devoted Health offers the Apple Watch as a benefit to its members (Farr, 2019). Devoted Health spokesman Kenneth Baer told CNBC's Christina Farr in 2019, "We are pleased that CMS agree that there is a wide variety of ways that older Americans can keep healthy, including fitness and nutrition classes, and activity monitoring devices such as the Apple Watch."[3]

And while other care providers such as senior living communities or home care agencies might not have the wealth or resources health insurers have, they too are considered coveted customers by agetech companies.

Validating both the problem and solution with real users—all types of users, building products that are not just beautiful but also usable and desirable and creating partnerships and solid business models—are some of the lessons that can be learned from successful agetech companies.

---

3   The **Centers for Medicare & Medicaid Services** is a federal agency that administers the nation's major healthcare programs including **Medicare, Medicaid**, and CHIP.

CHAPTER 8:

About the Role of Governments and NGOs

For quite a while, demographic projections have shown that the population is aging on an unprecedented scale. However, governments have been slow to respond and put in place policies that will position countries in a place where they manage the outcomes effectively. Although public funding and policies regarding long-term care and other elder care services have been in place, increasing the availability of these services takes years if not decades, since they mostly depend on human labor to operate.

When it comes to technology for older adults, it hasn't exactly been top of mind for governments. That's partially because policymakers and technologists live in two separate worlds and because policymakers themselves don't necessarily view older adults as potential users for cutting-edge technology (Schneier, 2019). This premise might have been true in the past, but that is no longer the case for the majority of the older population who's adopting all kinds of technology now more than ever.

Thankfully, the veil has been lifted from the eyes of policymakers, partially because of the COVID-19 pandemic and the uptick in demand for and adoption of tech by older adults. Suddenly, the digital divide was front and center, and it became painfully clear what the social costs of neglecting to connect our elders are.

Governments all over the world have taken it upon themselves to remedy the situation and rethink how to prepare for the aging of their populations at this large scale. This of course includes investments in infrastructure and technology. How are they doing it and how could it impact elder care providers and innovators in this space? Here are some examples for governments who are proactive about aging and tech.

THE UNITED STATES

The United States' economy is the largest economy on Earth, and it is also a target market for many agetech start-ups (Silver, 2020). So, when newly elected president Joe Biden's $2.3 trillion infrastructure plan includes $400 billion to support elder-care—everyone should be paying attention (Whitehouse.gov, 2021).

A fact sheet released by the White House about "The American Jobs Plan" in 2021 explains that this administration's aim is to "solidify the infrastructure of our care economy by creating jobs and raising wages and benefits for essential home care workers."

The administration's acknowledgment of the fact that the caregiving crisis is partially caused by low wages paid to caregivers and that elder care is infrastructure is incredibly important.

"President Biden believes more people should have the opportunity to receive care at home, in a supportive community, or from a loved one."

WHITE HOUSE 2021

The $400 billion will be used to do the following:

- Expand access to home and community-based services (HCBS) under Medicaid.
- Extend the longstanding Money Follows the Person (MFP) program that supports innovations in the delivery of long-term care.
- Support well-paying caregiving jobs that include benefits and the ability to collectively bargain (form unions).

Two other parts of the plan that could prove especially impactful for the advancement of agetech are the efforts to close the digital divide. The United States' government is tackling it from both angles: accessibility and affordability. It plans to do so in the following ways:

1. Invest $100 billion in internet infrastructure that will cover 100 percent of the country and make the internet more affordable.
2. Launch the Emergency Broadband Benefit (EBB) program, which will provide discounts for purchasing internet services to eligible households (FCC, 2021).

WHO COULD BENEFIT FROM THESE INVESTMENTS?

The significant $400 billion investment to expand home and community-based services and increase wages and benefits for caregivers could be like rocket fuel to an industry that's been steadily growing in recent years—the home care industry. *Home Health Care News'* Robert Holly reported in 2020 that the national spending on home health topped $100 billion in 2019. To put things in perspective, the entire home care market in the United States is estimated to be worth approximately $109.6 billion, and the number of businesses in it has grown 3.7 percent per year on average over the five years between 2016 and 2021 (Ibisworld, 2021).

Since many agetech companies view home care providers as either direct customers or important channels for distribution, growth in that industry increases market size for those companies.

Efforts to connect older Americans will effectively increase the market size available for agetech companies, since many of their products rely on having a high-speed internet connection in the home.

The US infrastructure bill also includes support for creating more caregiving jobs, with higher wages and more benefits with the purpose of solving the caregiving crisis, arguably the home care industry's biggest challenge. The caregiver shortage is a limiting factor for many older adults who prefer to age in place.

Home Healthcare News' Robert Holly reported in 2020 that caregiver turnover rates in the United States are higher than 60 percent. The average cost of replacing one caregiver is $2,600—these numbers don't leave providers with much time or financial resources to invest in technology (Kempton, 2018). If the bill is successful in improving the status of caregivers, this should go a long way toward reducing these costs and freeing up resources that providers could then invest in technology that could complement their existing, labor-intensive service.

If the demand for caregiving jobs increases, there will be a need to train lots of new hires. Companies like CareAcademy and InTheKnow offer online state-approved caregiver training that could train large numbers of caregivers at scale.

MONEY FOLLOWS THE PERSON

Entrepreneurs in agetech should view this bill as an opportunity from the Money Follows the Person (MFP) angle as well. According to Medicaid's website, one of the program's goals is to "increase the use of home and community-based services (HCBS) and reduce the use of institutionally-based services." Medicare and Medicaid have been covering remote patient monitoring and other home health technologies for years, allowing clinicians to extend their services into the home (Jacobson, 2017). Regulatory changes during the pandemic have finally made telehealth widely available.

MFP's emphasis is on aging in place, which today, more than ever, can be enabled with the use of multiple smart home technologies.

Telehealth services, home sensors and wearables for fall detection and wellness monitoring, VR solutions for rehabilitation, online fitness programs, and social robot and smart TV/screen-based solutions for loneliness and social isolation are just some examples for commercially available solutions, many of which are already reimbursable or covered by insurers. With the expansion of HCBS and MFP, it is highly likely that more government funding will be used to fund aging-in-place enabling tech.

NOTABLE EFFORTS TO SUPPORT AGETECH EXIST IN OTHER COUNTRIES AS WELL

In hindsight, we might view 2021 as a pivotal year when it comes to governments addressing agetech, tech in elder care, and the digital divide. While the United States is committed to making the largest monetary investments because of its size, other countries have also been making significant leaps forward.

The United Nations declared 2021 to 2030 the Decade of Healthy Aging: "A global collaboration, aligned with the last ten years of the Sustainable Development Goals, that brings together governments, civil society, international agencies, professionals, academia, the media, and the private sector to improve the lives of older people, their families, and the communities in which they live" (World Health Organization, 2021).

Some examples of countries that have taken major steps forward when it comes to promoting the use of technology for the benefit of the older population are Australia, Israel, Canada, and the United Kingdom.

Australia's aged-care reform places an emphasis on agetech, and the 2021 report from the Royal Commission into Aged Care Quality and Safety recommends to "implement an assistive technology and home modifications category within the aged care program that:

a. Provides goods, aids, equipment and services that promote a level of independence in daily living tasks and reduces risks to living safely at home
b. Includes the assistive technology, home modifications and hoarding and squalor service types from the Commonwealth Home Support Programme
c. Is grant funded."

There's also a recommendation that "the Australian Government should provide funding equal to 1.8 percent of the total Australian Government expenditure on aged care to the Aged Care Research and Innovation Fund each year."

Some of the countries that have high percentages of older adults in their population are in Europe, and in 2020, more than 20 percent of the EU population was aged sixty-five and over. The European Union is funding projects for technologies for aging well and has been doing so for quite some time. It has allocated €2 billion as part of the Horizon 2020 program to "keep older people active and independent for longer and support the development of new, safer, and more effective interventions."

In 2018, before it decided to part ways from the European Union, the United Kingdom appointed a minister for loneliness after government research found that "about 200,000 older people in Britain had not had a conversation with a friend or relative in more than a month" (Yeginsu, 2018). The United Kingdom also has a £98 million Health Ageing Industrial Strategy Challenge Fund that launched with the purpose to give older people five extra years of independent living by 2035 (Loughran, 2019). Business Secretary Greg Clark said: "As more people live longer, we must ensure people can live independently, with dignity and a good quality of life for longer by harnessing the best technological innovation and advances to help."

Israel, also known as "Start-up Nation," has its own Innovation Authority (IIA) that was created to "provide a variety of practical tools and funding platforms aimed at effectively addressing the dynamic and changing needs of the local and international innovation ecosystems" (Senor & Singer, 2009). In 2018, the IIA launched a collaboration with The Centre for Aging + Brain Health Innovation (CABHI) in Canada, which was intended "to accelerate the evaluation and adoption of innovative products and services that address the needs of aging adults and the challenges presented by an aging population." Israeli companies were invited to trial their solutions with leading elder care organizations from across Canada and receive up-to CAD $250,000 in funding.

The program was focused on the following innovation themes:

1. Aging in place
2. Caregiver support
3. Care coordination and navigation
4. Cognitive health
   (Israel Innovation Authority, 2021)

In 2020, a few months into the COVID-19 pandemic, Canadian Prime Minister Justin Trudeau announced "an investment of $1.75 billion to help connect Canadians to high-speed Internet across the country" with the goal of connecting all Canadians by 2030. Canada's government has been investing in aging innovation long before the pandemic, through organizations like CABHI and AGE-WELL. Canada's government also funds the Aging in Place Challenge program, which aims to reduce the number of older adults requiring nursing home care by 20 percent by 2031 (National Research Council Canada, 2021).

Dr. Alex Mihailidis is the principal investigator and a joint scientific director of AGE-WELL. His research disciplines include biomedical and biochemical engineering, computer science, geriatrics, and occupational therapy. He has published over 150 journal and conference papers in this field and co-authored/edited multiple books, including *Technology and Aging: Selected Papers from the 2007 International Conference on Technology and Aging*. Mihailidis is very active in the rehabilitation engineering profession and is currently the president for the Rehabilitation Engineering and Assistive Technology Society for North America (AGE-WELL, 2021).

As someone who is very much immersed in research and development of technologies that improve the lives of older adults and understands the intricacies of government funding for these efforts, Dr. Mihailidis is someone who understands how governments can improve their efforts on supporting agetech. He believes governments should "work much more closely with the academic field, the researchers, industry, etc. in these areas, networks like AGE-WELL and others around the world that really have their hand on the pulse of the sector itself."

He believes this is the way to "feed" governments with up-to-date information and make sure they're making decisions based on the most current developments in the field.

"They haven't realized the role technology can play in reducing their own health care costs."

DR. ALEX MIHAILIDIS

According to Dr. Mihailidis, the pandemic has proven to governments that technology can play a role in supporting the health of older adults, like when there was an uptick in adoption of telehealth services. He also believes it's important that the government plays a role in funding agetech start-ups as a means to attract international start-ups to open up headquarters in Canada and to serve the local population.

I agree with Dr. Mihailidis, and I am also quite optimistic. It appears that the United Nations declaration on a decade of healthy aging hasn't fallen on deaf ears, and governments around the world are making a conscious effort to improve the well-being and care of older adults by using technology. While it would have been better to have closed the digital divide and make sure all older adults have access to technology pre-pandemic, it's better late than never. At the time of writing these lines, there are still parts of the world that go in and out of lockdowns. Each and every older adult who has to shelter in place with no access to the outside world could benefit from having an internet connection, the devices to utilize it, and the digital literacy education to access online services and experiences.

HOW GOVERNMENTS AND NGOS CAN SUPPORT THE AGETECH REVOLUTION

Research and development of innovative technology to support older adults requires funding. One major thing that needs to happen is government funding for R&D. Research and academic institutes don't necessarily have the capability or even the will to commercialize technology. This type of research that isn't constrained by the need to find a business use case for a specific technology can later be used as a basis for future developments that will become a commercial product with real-world applications and a business model. The Israeli Ministry of Science and Technology created the "Science Accelerators" program to promote entrepreneurship and encourage entrepreneurship among researchers in academia and research institutes (Israeli Ministry of Science and Technology, 2021)

One example of academic research that was used to create a consumer product is the work of Dr. Cynthia Breazeal, who founded the Personal Robots Group at MIT's Media Lab. Dr. Breazeal later went on to build Jibo—the world's first home social robot, which kick-started development with a hugely successful crowdfunding campaign. Unfortunately, Jibo the company had to shut down operations in 2019, but Jibo's IP was acquired, and it is set to launch again as a health care and education product (Carman, 2020).

Additionally, governments can and should fund tech startups that are developing technology for specific challenges of aging in very early stages, when it's typically harder to get venture capital funding. This is similar to the type of work that the Israeli Innovation Authority is able to do: they support "innovative technological concepts at the pre-seed or initial R&D stages, transform their ideas into reality and reach significant fundable milestones."

There's also the option to provide funding on the consumer end. Since we already know the affordability of technology can be a barrier for adoption, and some technologies could promote the social determinants of health, it might be worth looking into subsidizing specific solutions.

In Israel, Personal Emergency Response Systems can be subsidized by government funding to those who receive government-funded home care. In some cases, it can also be covered by nonprofits. Why not provide subsidies for technologies that enable aging in place and tackle loneliness and social isolation? It could, if done correctly, have a positive return on investment and help reduce government spending on health care and nursing homes.

As always, I would have to say I'm optimistic. It seems that for the first time in modern history, governments recognize the importance of using technology to tackle social challenges, including the challenges of aging. Having many governments all over the world also making investments in R&D and providing funding to start-ups is a good stepping stone to accelerating innovation in aging.[4]

---

4   Parts of this chapter originally appeared on TheGerontechnologist.com

Future Opportunities

—

We live in extraordinary times. We carry in our pockets computers that are more powerful than the ones used by NASA during the Apollo mission, in which a human walked upon the face of the moon (Grossman, 2017). Scientists were able to develop a life-saving vaccine to a virus that caused a global pandemic in under a year, which is ten to fifteen times faster than it usually takes to develop a vaccine (Medical News Today, 2021). Meat that is grown in laboratories will soon be available for sale (Carrington, 2020). Flying cars could become available by 2023 (De Luce, 2020). In March 2021, Scott Galloway wrote in his *Medium* article "This Is the Best Time to Start a Business" that historically, post-crisis periods are extremely productive eras. In previous chapters, we discussed the challenges of aging and some tech-enabled solutions. In this chapter, I will make an attempt to point at what I believe are some interesting opportunities at the intersection of an aging world and the new normal a global pandemic has brought us.

According to Maslow's hierarchy of needs, human beings try to fulfill basic needs like food and shelter before they seek out a way to fulfill higher needs like self-actualization (Mcleod, 2018). Older adults are no different, and while the challenges of aging discussed in Chapter Four are a useful framework for sorting out the existing challenges, more opportunities are created by the aging of the population. Specifically, opportunities related to the higher needs on Maslow's pyramid—love and belonging, esteem needs, and self-fulfillment needs—are abound.

When I set out to write about future opportunities in agetech, there was one person I knew I had to get input from, and that is Mary Furlong (EdD). Furlong is the president and CEO of Mary Furlong and Associates, which was founded in 2003 and is a leading authority on the longevity marketplace. She has guided the business development and marketing strategies of leading corporations, emerging companies, and nonprofit organizations for more than thirty years. Furlong is also a pioneer in connecting older adults with technology. She founded nonprofit organization SeniorNet in 1986 and has a unique, thirty-plus-year perspective on where we were and where we could and should be headed. "When we were teaching seniors how to use computers, they were big computers, and I had one of the first Macintoshes—they called it a 'luggable.' When we first taught SeniorNet classes, people would pick up the controller as if it was a video game, so user interface design has come so far. . . . I went from the Mac 'luggable,' to now I'm going to get the iPhone 13 because of the camera quality; they became such an important part of our life."

When I asked Furlong what in her view has changed in the past thirty or so years when it comes to older adults and technology, she had this to say:

"Because of COVID, in the last eighteen months, everyone has thought about the technologies we need for older adults. SeniorNet net was focused on mitigating social isolation and empowering older adults with technology thirty years ago, and now finally, after thirty years or so, social isolation and loneliness have received so much attention."

UNRESOLVED CHALLENGES

Despite everything that has happened since March of 2020, the fundamental needs of human beings—food, shelter, human interaction, a sense of belonging, etc.—are the same. It was so easy to underestimate how much we needed human interaction when many of us were surrounded by people for most of our waking hours. And then, when a global pandemic hit, we found ourselves sheltering in place to protect our health. For many of us, the computer screen became our window to the world. Older adults were no exception, and their adoption of technology has skyrocketed during COVID-19. They not only purchased devices, but they adopted video chat, telehealth, and ecommerce more than ever before (AARP, 2021).

Part of this rush to adopt was driven by the fact that people were lonely and isolated, and the only safe human interaction that could occur was through a screen. Those who live alone and found themselves spending weeks or months by themselves felt it the most. For many older adults, that feeling was all too familiar.

ElliQ, the robotic social companion that I had the honor of participating in building, is one example of the way technology could be used to address an existing challenge of aging that suddenly everyone became aware of. But what about the challenges of aging that are just now beginning to scratch the surface in terms of public awareness, like the financial challenge? The cost of care? The caregiver crisis?

One concept that deserves more attention and is due for disruption is the concept of retirement. Retiring in your early sixties made sense when people were only expected to live for a few short years postretirement. These days, when life expectancy is at an all-time high, many people can expect to outlive their retirement savings. Many people can't afford to retire because they won't have enough to cover basic living expenses, let alone the added health care expenses that are expected in later years.

People lose more than an income stream when they retire. We define ourselves by what we do for a living, so when people retire, they can lose social status and their sense of purpose. For many of us, going to work is the main reason to get up in the morning, get dressed, go out of the house, and meet people. Going to work every day provides people with a daily dose of human interaction, which is a basic human need.

Postponing retirement is the logical solution, for those who are able to do so. The pandemic has already shaped how we work. I believe that the new, post-pandemic work world will provide a great opportunity to reshape workforce participation opportunities that are available to older adults.

Isn't it time to retire retirement?

EMPLOYMENT AND WORKFORCE PARTICIPATION

In her 2011 book *Happiness at Work: Maximizing Your Psychological Capital for Success*, author Jessica Pryce-Jones calculates that the average person will spend ninety thousand hours at work over a lifetime—that's over ten years total and 35 percent of waking hours for someone with a fifty-year working-life period. Work is only second to sleeping in the amount of time we spend on it during a lifetime. The way we work has been significantly disrupted since the onset of the COVID-19 pandemic. It still hasn't quite settled into a "new normal," and many believe this is an opportunity to reshape this part of our lives. In many ways—like forced early retirement and the lack of quality job opportunities for older workers, the current job market is failing older adults, and this presents opportunities for innovation.

As discussed in previous chapters, demographic changes mean that the dependency ratio is changing in many countries. The dependency ratio is a measure that compares the number of "dependent" individuals who are either below or above working age, with the number of working-age people. This ratio is expected to rise in all G20 countries in the next decades (Rouzet et al., 2019). This could lead to decreased economic growth (Santacreu, 2016).

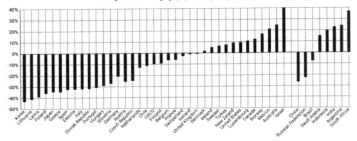

The working-age population will decline in a large number of OECD countries
Change in the working age population (20-64), 2020-2060

Source: United Nations World Population Prospects: The 2019 Revision

Source: OECD (2019), "The working-age population will decline in a large number of OECD countries: Change in the working age population (20-64), 2020-2060", in *Pensions at a Glance 2019: OECD and G20 Indicators*, OECD Publishing, Paris, https://doi.org/10.1787/0f86bb5f-en.

"The problem won't just be a lack of bodies. Skills, knowledge, experience, and relationships walk out the door every time somebody retires—and they take time and money to replace."

(DYCHTWALD ET AL., 2004)

Japan, which has one of the oldest populations in the world with almost a third of people over the age of sixty-five, also has some of the lowest GDP growth rates among G7 countries (Edmond, 2019).

The global COVID-19 pandemic has cleared offices around the world almost overnight as workers made a swift shift to working from home. With many people working remotely until further notice, companies were more open to hiring remote talent, which opened up opportunities for knowledge workers to live far away from urban employment areas. This shift has also been beneficial for people with disabilities who often have to take accessibility into account when choosing their place of work.

Older workers have struggled to find work since before the pandemic. After losing their position, workers over the age of fifty take longer to find a new one (Weller, 2019). This reality created an opportunity for services that specialize in helping older job seekers find a position, like Maturious. The pandemic has caused employment for older workers to drop significantly: the decline in the employment rate for those over fifty has been much more significant, and many of those who work had to take pay cuts (Cominetti, 2021). Millions of older workers were forced into early retirement (Marcus, 2021). The challenges older workers face in the job market might be a contributing factor to the rise of "boomer-preneurs": Baby Boomers who are starting new businesses. When I asked Furlong about this, she said: "I love that trend, and I think it's only going to continue. There will be many people who say, 'Why not take the acknowledge and experience and wisdom that we have, and use that to build a business, whether at the local level or globally through the internet,' [and] I think we're going to throw out the word retirement and look for a new word that has to do with life-long learning and contribution and being builders."

REINVENTION

Dr. Rachel Korazim is a seventy-five-year-old educator who has reinvented herself twice in the thirteen years that have passed since she retired from her work at The Jewish Agency for Israel. The first time was to become an entrepreneur traveling the world, teaching the Jewish diaspora about Israeli culture through poetry, and the second time was transforming her business from being in-person to online.

Throughout her career, as she rose up the ranks in the world of education, she got further and further away from doing what she was most passionate about—teaching poetry. And so, when she had to reinvent herself at the age of sixty-two (the current age in which women in Israel can retire), she knew in the next chapter of her career, she would do just that.

The niche she chose was to teach Israeli poetry in English. In Israel, she would teach at places like the Pardes Institute or Shalom Hartman Institute—two institutes that are located in Jerusalem and work primarily with English-speaking audiences. Abroad, she would mostly work with Jewish communities in the United States. She would go on a three-week tour multiple times a year. "My audience is spread from Miami to Boston and from Seattle to San Diego, but there are also communities in Arizona and Texas that I visit," she said.

When you go on a multicity tour, you have to handle flights, hotels, car transport, and communication with numerous people from different organizations that have booked you. Handling the logistics of booking multiple tours, one to two years in advance is a lot of work. Add to that the fact that the United States is a good ten-to-fifteen-hour flight from

Israel (depending on whether you're headed to the East or West Coast) and that you have to be on top of your game when giving talks, it's quite the challenge for a one-person operation. When I commented on it, Dr. Korazim had this to say: "These things aren't done with pen and paper. You have to learn how to do them on the computer, and if I hadn't acquired those tools for myself, it would have been significantly more difficult for me."

Suddenly, a global pandemic hits, and months upon months of speaking engagements get canceled one after the other. "You have a plan for the next one to two years and everything crashes down. . . . I woke up one morning, opened my phone and saw a video from Italy of people singing on their balconies. . . . I got out of bed and said, half-frustrated to [my spouse] Yossi, 'I can't sing like the Italians, but I can teach poetry.'"

She walked straight from the bedroom to the workroom, turned on her laptop, and logged into Facebook to write a post inviting people to join an online poetry class. Around forty people joined that first class, which turned into a regular class happening five times a week. Soon, a community emerged from those classes—this was a side effect Dr. Korazim could not have foreseen.

"It was quite obvious, from the second class onward, that it didn't matter what I would teach. People wanted to be there together. They wanted to see each other's faces, they wanted to have a conversation with one another, they wanted to share their stories, their fears. . . . Some of these people lived so far away from their family and kids. The community center was shut down, the mall was shut down, the synagogue was shut down—and they had my classes every day, and they would look forward to them."

One day, Dr. Korazim got an email from Esther, a community member, saying she had to miss class for a few days because her mother had passed away. The woman lived in Canada, and her mother had been living in a senior living community in Israel. Because of the COVID-19 travel restrictions, she never got to say goodbye.

One of the rituals in Jewish tradition is welcoming someone back into the community after the "Shiva," the seven days of mourning after a loved one's passing.

"Would you like us to welcome you back into the community?" Dr. Korazim asked.

"Will you do that for me?" Esther responded.

And so, the community conducted a virtual ritual, hosted by Linda, a retired Rabbi from Brooklyn, to welcome Esther back into the community. When asked about their most memorable moments in the community's one-year anniversary, everyone's response was: "Apart from Esther's welcome back ceremony, right?" It was obvious that this was the most memorable moment of all.

Dr. Korazim's reinvention story and the community she had built in the midst of a global pandemic, are great examples of how technology can enable us to reinvent ourselves at any age and how it can be used to bring people from all over the world together, to connect and share meaningful memorable moments together.

RESKILL VS. RETIRE

Some might think exiting the workforce earlier than expected is not that big of a deal, but the reality is many people don't have enough saved for retirement, and each year of employment is critical to their ability to sustain themselves for the rest of their lives post retirement, which could be twenty or thirty years, depending on life expectancy and age of retirement. Even for those who do have enough savings, a forced exit from the workforce before they can withdraw from their retirement funds might put them in a tough spot, not to mention the other benefits stemming from staying employed, like daily social engagement and a sense of purpose.

There's a lost decade of employment—between the age workers are either forced to retire or forced to settle for low-paying, part-time jobs—and their actual age of retirement. If making sure that those who are able and willing to keep working until or past retirement age is good for the economy and companies' bottom lines and creates clear incentives for individuals to postpone retirement, what can be done to make this a reality?

In her 2021 article for the *World Economic Forum*, Sofiat Akinola suggests some changes, such as "providing more flexible work options and retirement options, promoting lifelong learning, and creating age-friendly workplaces and overall better working conditions."

She also states that "retraining, reskilling and upskilling improves employability of all workers throughout their lives."

This falls in line with the need to reskill many workers who will soon be replaced by automation, which is another trend that has been accelerated by the pandemic.

The reskilling of millions of older workers could be part of a bigger play of reskilling tens of millions of workers in occupations that will soon be obsolete. Not all employers invest in ongoing education for their existing employees, and even those who do might be reluctant to offer training opportunities to older workers (The Center for Research into the Older Workforce). Oftentimes, individuals who want to keep learning throughout their careers have to take matters into their own hands. This has become easier in recent years with the proliferation of online learning platforms like Udemy and Skillshare.

What about workers who don't realize their skills are outdated? Those who don't get to amass valuable knowledge on the job? And those who will need to change occupations and industries altogether?

Can existing solutions reskill them at a scale that's big enough to change the trajectories of entire economies like Japan? Most importantly, when thinking of older workers, how can we make sure they find their way back into the workforce after reskilling?

LEISURE AND LIFELONG LEARNING

Post retirement, people suddenly have a lot of time on their hands. When I asked Furlong what she would do if I were to give her $1 million to start a new venture, she said: "I still think the media space is underserved, so I would probably do something similar to what I had done in Third AgeMedia (the media company she started in 1996). I would focus on vitality, I would focus on financial resilience, and I would focus on adventure and travel, and maybe a bit of romance. Because I think that this is going to be a group that sees more adventure ahead than they did behind them."

Travel tech, education tech, online dating, and social networks are all big industries worth billions of dollars. Very few companies that have tech-enabled offerings in these spaces actively pursue older adults as potential users, despite the economic potential.

Pre-pandemic, Airbnb reported people over the age of sixty are "our fastest growing age group, both as hosts and guests" (Greeley, 2018). The cruise industry has always attracted many older clients who can afford to take long leisurely cruises (Ship-Technology, 2019). Both Airbnb and the cruise industry have taken a hit during the COVID-19 pandemic, but this doesn't mean older adults aren't interested in traveling anymore. There's an opportunity right now to create new types of travel, pandemic-friendly types, that will enable older adults to explore the world safely. Who will take on this challenge?

As for online learning, there's so much to choose from both in terms of platforms (Udemy, Skillshare, and Coursera, just to name a few) and in terms of the types of things you can learn online. If you're looking to learn specific skills to get ahead in your career, or if there's a specific hobby you want to get into, it's relatively easy to find the right online education—you already know what you're looking for.

But what about older adults who just know they want to keep on educating themselves for personal development, just for the sake of learning? Finding the right online course among millions of online courses is a needle in a haystack.

"There's an ongoing need for technology, education, and continued learning. And that's only going to grow, because it's going to keep you in the workforce. It's gonna allow you to create your own business, if you like"

MARY FURLONG, 2021

ROBOTS AND HOME AUTOMATION TO SOLVE THE CAREGIVER CRISIS

Decades from now, when I'm older, I fully expect to have a live-in robot to help me with my ADLs and household maintenance, if I should require assistance. I know this might seem like a dystopian future for some of you reading this, but the reality is we simply won't have enough human caregivers to care for the two billion older adults who are expected to live on this planet in 2050. These days, a major part of the caregiver's role revolves around household chores and maintenance, grocery shopping, cooking, cleaning, and, of course, helping the care recipient with ADLs like bathing and toileting. All of this hard work leaves them with little time to do the things that are uniquely human, like providing comfort and companionship. I believe that families should get to spend the last few years in the lives of an older loved one spending quality time together, having meaningful conversations, and sharing joyful experiences rather than in scheduling appointments, running errands, sorting pills, cooking, and cleaning.

We can already automate some of the tasks currently being done by human caregivers: sorting out pills, creating reminders, and dispensing the right pill at the right time can be done by a robotic pill dispenser like Black+Decker's Pria. Other tasks, like scheduling appointments, will be available in the near future. Google's Duplex allows Google Assistant to make calls for you and schedule appointments with local businesses and is currently being rolled out to select geographies and devices (Tillman, 2021). There's plenty more we can and should automate in order to give family caregivers and professional caregivers the opportunity to have more quality time with their older loved one/care recipient.

We could build home robots that could handle household maintenance and help with ADLs. We could develop ambient technology that's embedded in our homes, able to anticipate our needs, and provide us with instant solutions. There's more on this in the next chapter.

"It started out as a quiet march, now it's a global chorus of talent."

MARY FURLONG, 2021

The above quote was taken from the conversation I had with Furlong in the final weeks of writing this book. She told me that she is inspired by the entrepreneurs she meets in this ecosystem: "Their energy and idealism is what inspires me. Their success, knowing that they have collectively changed the lives of so many older adults—that is the magic of the orchestra of talent that has contributed to the longevity revolution. It started out as a quiet march, and now it has become a very large universe of people who are chipping away at the problem of trying to use technology to make life better for older people. That's a global chorus of talent."

During the COVID-19 pandemic, tech adoption rates among older adults skyrocketed. E-commerce, telehealth, and video chats also got a huge boost.

Will older adults go back to their old habits when the pandemic is over? I don't think so.

In 2030, which is just around the corner, we could have a world that isn't just tech driven but also age inclusive. We could have a workforce that enables workers of all ages to contribute their skills and talent. We could automate our homes and build robots that will enable more people to age in place. We could definitely utilize technology to help each and every older adult achieve their goals and live their best life.

CHAPTER 10:

A Possible Future

———

March 24, 2035, was supposed to be the day Ethel died. It had started like any other Friday, with the aroma of coffee spreading throughout Ethel's apartment. Being retired, she had the luxury of waking up any time she liked, but this was no ordinary Friday. Ethel owned a state-of-the-art smart coffee maker that was able to connect to smart home systems that tracked sleep cycles and quality, with the purpose of waking its owners when they were fully rested. Trusting the machine to wake her up toward the end of her sleep cycle so she would feel well rested, Ethel set it to start brewing at dawn. That coffee maker was a gift from her granddaughter Helia, who knew how much Ethel liked coffee (and sleeping) and thought it would be perfect for her. She was right.

Ethel had been living alone for almost five years since her partner passed away. They were together for fifty-one years, and even though Ethel was living alone for all this time, it still felt strange to her to wake up in bed all alone. The stillness was another thing. Her old home had gotten too big and quiet, so she decided to downsize. Two years prior, she moved into a nice modern apartment in one of those buildings that had everything "smart" in them. The brochure promised her a living experience she couldn't even have imagined not long ago. She still remembered a time when fax machines and rotator phones were considered advanced technology. Not even in her wildest dreams could she have imagined a home that was equipped with so many sensors that it could cater to her every whim without her needing to lift a finger to do even the slightest amount of maintenance or housework. It was just what she needed. After all, she was eighty-five years of age already, and at her age, she did not want to waste her time on changing light bulbs or vacuuming. Deciding to move was a no-brainer.

As the aroma of coffee reached Ethel's bedroom, gently nudging her awake, the shutters started opening slowly, letting more and more light in. She opened her eyes and stretched her limbs while lying down and then said, "Good morning, Casa!"

"Good morning, Ethel," Casa replied. "Should I play the news?"

"Yes, thank you, Casa," said Ethel.

As Ethel got up from her bed and moved toward the kitchen to get her coffee, the voice of the news anchor followed her around, playing from a different set of speakers each time. Ethel liked her morning routine, which always included drinking coffee and listening to the news on her balcony overlooking the bay. After decades of living in the suburbs, she enjoyed the hustle and bustle of the big city. She was delighted to rediscover everything the city had to offer. Now that she was widowed, being busy was a blessing, and in the city, she didn't have to be alone. The theater, the park, the restaurants, her gym—everything was within walking distance or a short car ride away from her apartment building. The best part was she was close to the people she loved. Most of her children and grandchildren lived nearby and many of her friends moved back into the city as well. It seemed everyone was downsizing these days.

As she finished drinking her coffee, Casa asked whether she would like to start the day with an invigorating yoga routine. "Sure, why not," she replied, thinking to herself that it not only helps loosen up those stiff joints in the morning, but also that she always felt an inch taller after yoga practice. When she walked into her closet, her yoga clothes were extended on a cloth hanger, and as she got back to the balcony, her yoga mat was handed to her by Nietzsche, her robotic helper. As she walked onto the mat to start her practice, a holographic yoga teacher appeared. "Good morning, Ethel," the hologram said. "We will start today's practice with a few rounds of the sun salutation. Ready?"

Ethel started her routine, and with her newly installed 360-degree fitness cameras, the teacher was able to view Ethel from every angle, offering guidance on how to adjust her form and perform the routine in the best way possible. As she was doing her sun salutations, she felt her right knee was stiffer than usual. "Starting the day with yoga was a good call," she thought, although a slight suspicion crept into her mind that it might be a sign that this knee would soon need replacing too.

As her morning yoga routine came to an end, she could smell the scent of something cooking in the kitchen. Although she liked being in control, she had set her meal preferences on Casa to "Surprise me," thinking meals don't have to be boring. She had recently asked it to download a new recipe for frittata and from the smell of it, she guessed that it had decided to order the ingredients and make it for her this morning. "Casa, what's on my schedule for today?" Ethel asked on her way to the kitchen.

"Today is Helia's birthday, and you are meeting her for lunch at Sorbillo's at noon."

Ethel remembered she had ordered a special birthday gift for Helia, a rare book that was supposed to arrive yesterday. "Casa, did Helia's gift arrive yet?" Ethel really wanted to see the expression on Helia's face when she unwrapped it.

"Let me check," Casa replied. A few seconds had passed, and Ethel started eating and daydreaming about how happy Helia would be when she received her gift.

"Ethel?" Casa said with a tone of worry in its voice, as if it was afraid to tell Ethel the bad news. "I'm afraid that package got sent to the wrong address."

Ethel stopped chewing.

"Where is it?" she asked.

"It's in a bookstore on W 132nd Street," Casa replied.

"What time do they open?" Ethel asked.

"Ten a.m.," said Casa.

"Well, this isn't what I had planned for this morning, but I guess I could head over there to pick it up myself." Ethel thought to herself.

"Casa, can you reach out to them and tell them I'll be coming to pick it up?"

Within a second, Casa replied, "I just spoke to their service bot, and it promised it would hold the book for you."

"Thank you, Casa, and please have a car pick me up when I finish getting ready."

"You got it, Ethel!"

She disliked speaking to those service bots. They had low-grade emotional artificial intelligence and got her irritated every time. It was such a relief to have Casa deal with them for her. Casa was three generations more advanced and was basically built to keep her happy, and since it was speaking the same language as they were, it got those service bots to do exactly what Ethel needed them to do, every time. Explaining to them that a book was accidentally shipped to them and that she would be coming to pick it up would have taken Ethel a few minutes. They were not very good at dealing with unusual inquiries, and having a book shipped to them by accident was certainly unusual. For Casa, it only took a second.

"Thank you Casa, that was delightful," Ethel said when she finished eating breakfast, heading over toward the shower, knowing that the dishes will be sparkling clean in a few minutes, and there was nothing she had to do to get it done.

"You're very welcome, Ethel," Casa replied.

As she stepped into the shower, she clapped her hands twice and the water started running at exactly the right temperature for a postworkout shower. Beethoven's Moonlight Sonata started playing on the speakers. "Casa is feeling the drama this morning," Ethel thought to herself and smiled. As she started putting on her favorite facial cream, she noticed it was about to run out. "Casa, order this face cream," she asked and held the jar steady in front of the mirror for scanning. "Face cream added to your shopping basket. How soon would you like it delivered?" Casa asked. She looked into the jar and replied, "Next week would be okay."

Her wardrobe extended to her a clothing rack with weather appropriate clothing. "It's finally getting warmer," Ethel noted to herself and picked up a nice pair of jeans and a cotton T-shirt. Her friends liked to say that she didn't dress her age, but what does that even mean? She smiled at her own reflection in the mirror and put some lipstick on before heading toward the door just as Casa said, "Your car is outside, Ethel!"

The shutters facing the south in her apartment slowly closed so that the house wouldn't get too hot in Ethel's absence. She then closed the door behind her and entered the elevator. "I have plenty of time to get across town and back in time for lunch with Helia," Ethel thought. As she stepped into the car, the holographic driver turned its head to greet her. She silenced it with a hand gesture. "They didn't quite get these right," she thought to herself. The look in their "eyes" always gave her the creeps, but they say that some people feel safer seeing "someone" behind the wheel, even though they know it's a self-driving car. These cars were a lifesaver to her when she had knee replacement surgery last year, so it had been quite some time since she took out her vintage Chevrolet out of the garage to drive herself anywhere. The only reason she hadn't sold it yet is because it was a collector's item, and she thought it would be a nice inheritance for her son, who was a car fanatic.

Chopin's Nocturne in E-flat major was playing on the speakers, but Ethel felt like listening to something a little more upbeat. "Auto, play some Beyoncé please," she requested. "Single Ladies" started playing and Ethel instantly felt more cheerful. "That's better," she thought to herself and reclined back in her seat, knowing that she had at least twenty minutes to get across town, if the traffic was light. She had always preferred being a passenger, letting someone else do the driving and spending her time in the car staring out the window and letting her thoughts wander.

Driving always felt like a chore, and when humans were behind the wheel, it was also stressful. Humans are unpredictable and you never know whether the drivers around you had one too many drinks before getting behind the wheel. It was a beautiful spring day outside, and Ethel let her thoughts carry her away. She started reminiscing about that spring, twenty-nine years ago, when Heila was born. She was such a delightful child, always smiling, always happy to spend time with her grandparents. Ethel was there when Heila took her first steps, said her first words, and as she grew up and went to kindergarten, school, university. . . . The years went by so fast; it was almost unbelievable that Heila was almost thirty years old and had her own consulting business. Ethel was incredibly proud of her and kept telling all her friends about it.

"We have arrived," Auto announced, and Ethel's mind was pulled back into reality. "Wait for me here, please," Ethel told Auto as she exited the car. There was a "no stopping" sign where the bookstore was, so she had to cross the street to get there. It was a busy four-way intersection, with four lanes in each direction, and Ethel was grateful for the cover of clouds as she waited for the traffic light to turn green. Suddenly, her earpiece rang. "Incoming call from Dr. Fitz," the earpiece said. "I better get this," Ethel thought to herself, gently touching her earpiece with her index finger.

"Hello?" Ethel said.

"Hi Ethel, it's Dr. Fitz, how are you?" asked Dr. Fitz.

"I'm doing well, how are you, doctor?" Ethel replied.

"Great, I'm happy to hear you are doing well," said Dr. Fitz. "Remember that when we replaced your left knee last year, I told you that since your right knee is just as old, it will eventually need replacing too?"

"Yes, I do," Ethel replied as she realized that the pedestrian traffic light had already turned green. She started crossing the intersection, and two construction workers who were hauling a huge, reflective window walked beside her, hiding her from view of upcoming traffic.

"Well, it could be nothing, but our system predicts that your right knee will need a replacement sometime during the next quarter. I'd like you to come in so we can scan it and get a closer look," said Dr. Fitz.

"Actually, Doctor—" said Ethel, but before she could complete the sentence, she noticed a vintage car speeding toward her. This one was driven by a real-life human being, and he was blinded by the sun glare reflecting from the window carried by the workers, who had since passed Ethel. The pedestrian traffic light had turned red, and as the car light turned green, the driver headed toward the intersection, full speed ahead, not noticing Ethel.

"Is this the way it ends?" she thought to herself as she instinctively leapt forward, using every muscle in her body to propel herself. It was a split-second decision that wasn't even conscious—whether to leap forward or backward to avoid the car. Even if the driver noticed her now and applied the brakes, they had no way of stopping the car in time to avoid hitting her, so the car had to sway in one direction. Which direction would it be?

The construction workers had noticed the speeding car and started waving their arms and yelling, "Stop the car! Stop the car!" One of them extended a hand to Ethel, grabbing her upper arm and pulling her toward them in an effort to help her move out of the way quicker, but it was too late, the car was less than thirty feet away from them, traveling at thirty miles per hour. There was no way the driver could do anything to avoid a head-on collision, but at the last split-second, he noticed Ethel and initiated an emergency brake. The car swayed to the right, narrowly avoiding Ethel and racing past the intersection, coming to a halt after it. The driver, a fit, forty-something-year-old man in a suit, got out of the car and ran toward Ethel.

"Lady, are you okay?" he asked as a visibly shaken Ethel rested on the sidewalk, trying to steady herself with some support from the construction worker, right next to the spot where she nearly lost her life.

"I think so," Ethel replied, her voice shaking and her breath shallow. Her mouth was dry, and she could feel her heart beating in her chest. It felt good to know it was still beating.

"I didn't see you; I don't know what happened! I think the sun got in my eye," the driver said. "I'm sure glad I got the new 'autonomous' system installed in it last week. I'd hate to think what would have happened had it not taken control of the steering wheel at the last second."

"Ethel, are you there? What happened?" Dr. Fitz's voice echoed in Ethel's earpiece again. "I'll tell you when we meet," Ethel responded. "Got to go, sorry" she said, touching her earpiece again to hang up.

"Listen I'm sorry, my wife tells me I shouldn't be driving anymore, that it isn't safe. Perhaps I should listen to her and only take this car out on the weekends," the driver said.

"She makes a valid point," Ethel said. "The important thing is that I'm okay and no one got hurt, but you might not be so lucky next time."

"You're probably right," the driver said. "Can I do anything to make it up to you? You must have been frightened, and I feel so bad knowing I probably ruined your day."

"Well, you could get into the Auto that's waiting for me across the street and get it to do a U-turn and wait in front of the bookstore when I come out in a few minutes," Ethel replied.

"Sure thing, it's the least I could do!" the driver said.

And so he did.

Ethel got into the bookstore and came out with the book wrapped in gift wrap in no time and then got back into the car.

"Take me to Sorbillo's, please," she told Auto.

"On the way! It would take approximately thirty-five minutes because of traffic," Auto replied.

"Thank you, Auto."

The rest of Ethel's day was blissfully uneventful.

***

This story may be science fiction today, but it provides a vivid example of how tech could improve the lives of older adults, not just by enabling aging in place and making homes easier to maintain but also by saving lives. The rates of pedestrian deaths in motor vehicle crashes per one hundred thousand people are highest for people aged seventy and older (IIHS, 2019).

Epilogue

The idea of writing a book crept into my mind about two years ago. At first, I didn't feel I had enough experience to write one. I had always thought I needed to get just one more hole in my belt to write a book, that starting my own company and becoming a legit tech entrepreneur would somehow magically transform me into an author that had something meaningful to say.

What I failed to realize was this was classic imposter syndrome, and in fact, I had plenty to say. Not just that, but people wanted to listen, read my blog, reach out to me for advice, and fly me to other continents just to sit in a panel or give a keynote. I was being perceived as a thought leader in my industry before I had realized that had happened.

So, one random Wednesday, in a hot Israeli summer, I was standing in the new office of my start-up company (we had moved four times in one year because of growth and a global pandemic), and then it hit me. I had to write this book.

In that aha moment, it became crystal clear I had the topic in mind all along. I would like to write about the agetech revolution. I have, after all, been writing about this for the past four years to a loyal audience that grew to be tens of thousands of people. At this stage in the life of the agetech ecosystem, it kind of feels as if we all share a secret about this huge thing that's on the verge of exploding (the good kind of exploding).

I realized this was too important to keep a secret, so I wanted to share it with the world. I did not, however, have all my ducks in a row. Writing a book isn't as simple as writing an article, and even writing an article can be f***ing hard at times.

What am I **actually** going to write about? Who will want to read this book? Where do I even begin? Luckily, I had people to reach out to in order to help me answer those questions.

I always say one of the greatest perks of having online visibility is that I get to have many interesting conversations with interesting people from all over the world. In the two years that have passed since the idea of writing a book first came to mind, I have had two fateful conversations that were critical milestones on my path to actually going through with it and publishing.

One of those conversations was with someone who later became a friend and a mentor, Dr. Jon Younger. I met Jon via Zoom back in 2019, soon after I was featured in *Forbes* for being part of *Next Avenue's* annual *influencers in aging* list.

Jon had wanted to interview me for a *Forbes* column about the freelance revolution and the future of work. I didn't know what to expect. It turned out Jon was (and still is) a fascinating human being—and also, as they say in Yiddish, a *mensch*.

We kept corresponding and catching up now and then via Zoom after that interview. Jon was one of the first people I told about my intention to write this book, and he was positive, encouraging, and insightful as ever. Jon is also one of a few people who agreed to read the full manuscript prior to publication and provide feedback, for which I am eternally grateful.

The other pivotal conversation happened almost by chance. I had received a book in the mail one day that carried the title *Digital Remains*, written by author Jarred Harrington. Jarred is a funeral director who noticed an interesting phenomenon during his work with families of people who were recently deceased. When people die, they leave behind them not just physical or monetary assets, but also digital assets. He decided to write a book about death, social media, and technology.

The book arrived at the same Tel Aviv office in which I had my aha moment. It contained a note from the author. I reached out to him via email and thanked him. A few months later, I realized I had never heard back (I blame spam filters) and decided to reach out to him via social media instead. Success!

As it turned out, it was Jarred's wife, JoHannah, who had sent me a copy of his book. Jarred was a second-time author. When I asked him how he got to writing his book he told me about the Creator Institute Program at Georgetown University. It was about a week before the application deadline, so my timing was spot on.

I feel like I've had this book in me for a while and needed a gentle "nudge" from the universe to write it. That nudge came in the form of an aha moment, followed by advice from fellow authors Jon Younger and Jarred Harrington. I am full of gratitude to the universe for bringing those signs to my path just when I needed them.

I hope that after reading this book, you are convinced we must use technology to tackle the challenges of aging, that we can't exclude older adults from the digital transformation our society is going through, and that tech companies, older adults, family caregivers, elder care providers, governments, and NGOs can all contribute to this revolution. I also hope that you too are optimistic about the future and confident that we do have the means to make a positive change in the way aging is experienced today using technology so future generations will get to live a life that's full of purpose, social engagement, and little to no household maintenance :)

I invite anyone reading this to continue to follow my work on TheGerontechnologist.com—there's a mailing list to which you can subscribe. I also invite you to connect with me on LinkedIn and Twitter. I would love to hear your thoughts on *The AgeTech Revolution*.

"To infinity and beyond!"

(BUZZ LIGHTYEAR, DISNEY PIXAR'S TOY STORY, 2009)

Sincerely,
Keren Etkin

Acknowledgments

———

Even in my wildest dreams, I could not have imagined the outpour of support I received from family, friends, colleagues, and the agetech community while writing this book.

I am truly grateful to have had so many wonderful people alongside me on this journey.

I would like to start by thanking my partner, Avishag, who encouraged me to spread my wings and make the career transition into tech, who continues to support me in anything I do—none of this could have happened without you. You are my rock and my sunshine.

To my family—my parents who are my biggest supporters, who allowed me to venture out into the world knowing I have someone to rely on, to my amazing sisters, to my grandparents who helped raise me and to everyone in my extended family who supported me on this journey—I am lucky to have you.

To my friend and mentor Dr. Idit Harel—you are an inspiration and a force of good in this world. Thank you for being part of my life and for your wonderful advice.

To Adam Etzion—thank you for your sharp insights and for supporting me as a writer while also preparing delicious food.

To Dr. Jon Younger—thank you for your wise words of encouragement and for your friendship.

To Nil Meral, you're the best, and I'm thankful to the universe for making our paths cross.

Thank you to everyone who shared their story with me, connected me with interviewees, and provided quotes: George Lorenzo, Dr. Rachel Korazim, Amanda Rees, Esther Hershcovich, Jake Rothstein, Stephen Golant, Alex Mihailidis, Davis Park, Sherry Baker, Don Norman, Mary Furlong, Jon Warner, Marie and Annmaree Desmond, Pat Whitty, Chip Connelly, Phil Davis, Burn Evans, Nathan and Barbara Firer, Jon and June Harrow, Phoebe Innes-Wilson, Arthur Bretschneider, and Erica Powell. Extra special thanks to Margaret Polanyi from AGE-WELL.

Thanks to Liz Miller, Richard Caro, Michael Phillips and everyone who referred me to valuable resources.

Thank you to everyone who reviewed the manuscript and gave me their honest feedback!

Jon Younger, Adam Etzion, Shani Ganiel, Frona and Ted Kahn, Jon Warner, Stephen Johnston, Idit Harel, Jonathan, and Miri Erez.

Thanks to those who are promoting aging innovation on Clubhouse and helped me spread the word about the Indiegogo campaign: Stefano Solerio, Linda Sherman, Michael Phillips, and Steve Ewell.

To Jarred Harrington and his wife JoHanna—thank you for the inspiration and for referring me to the Creator Institute.

To everyone in my author community who supported the presale campaign:

Drapin, Ryo Haruyama, Raheema Hemraj, Mike Billings, lital josifov, Takahiro Tanaka, David Cohen, Joachim Kurt Cerny, Peer Bentzen, Michael Hesser, Jeremy Alan Depp, Tracy Moore, Noam Stern, Tostor74, Barbara Croyle, Alexandre Faure, Don Kramer, Ilya Mitin, Carmela Mayer, Jim Murphy, Jacqueline A. Silverman, Jess Nachlas, Patricia Giramma, Xiang Lan, Chris Novosielski, Jeffrey Gray, Matt Golden @ MapHabit, Terence Ronson, Eunice Yang, Kerry Burnight, Yonah Liben, Yaniv Stern, Dr. Mario Geissler, Joel Shapira, Stephen Farber, Jan-Willem Callebaut, Michal Herz, Sarah Waxman, Sensorscall, Or Azulay, Michal Halperin Ben Zvi, Sara kyle, Vishal Asthana, Guillaume Lasalle, Shubhada Saxena, Ken Saitow, Dan-ya Shwartz, Jon Charles Warner, Liz Loewy/EverSafe, Erin McCune, Betsy Jones, Shai Granot, Mel Barsky/ CABHI, Jaison Jose, Torsten Dilba, Grace Andruszkiewicz, Shay Zweig, Leanne June, Roger Tjong, Stephen Ewell, Danny Lauber, David Lethbridge, Kevin Kelley,

Tracy Chadwell, Omer Briller, Anna Dubrovsky-Gaupp, Ron Beleno, Oded Hershkovitz, Gracyn Robinson, Ashley Bullers, Michael Skaff, Tami Amir, Nicholas Freeman, Moran Yossef, Ido Biran, Erika Walker, Sara Gonçalves, Gerold Manthey, Adi Uznik, Michal Derfner, Lital Eshel, David J Fryman, Tanya Denisyuk, Adi Scope Karmon, Dani Waxman, Jeff Weiss, Robi Ferentz, Frona Kahn, David Young, Frances West, Stephen Golant, Max Zamkow, Ashish Mudgal, Ira sobel, Matan Koby, Fae McKenna, Otto J Janke, Heidi Culbertson, Hila Etkin, Jonathan Younger, Haim and Tali Etkin, Miri erez, Menachem Etkin, Irit Shubert, Adva Levin, Danny Hadar, Moti Karmona, Yuval Shubert, Adam Etzion, Idit Harel, Lior Etkin, Dor Skuler, Ella Krutokop and Dafna Pressler from Intuition Robotics.

Extra special thanks to Professor Eric Koester and his team from the Creator Institute for creating this awesome program, and thanks to my editors Trisha Giramma and Rebecca Bruckenstein for being patient and holding my hand through this process. Thanks to the entire team at New Degree Press.

Appendix

———

INTRODUCTION

AARP Public Policy Institute. "The Aging of the Baby Boom and the Growing Care Gap: A Look at Future Declines in the Availability of Family Caregivers." *Insight on the Issues*, number 85: August 2013. Accessed September 21, 2021. https://www.aarp.org/content/dam/aarp/research/public_policy_institute/ltc/2013/baby-boom-and-the-growing-care-gap-in-brief-AARP-ppi-ltc.pdf.

Coughlin, Joseph F. and Luke Yoquinto. "Technology for Older People Doesn't Have to Be Ugly." *Wall Street Journal*, October 14, 2018. https://www.wsj.com/articles/technology-for-older-people-doesnt-have-to-be-ugly-1539546423.

Edmond, Charlotte. "Elderly People Make Up a Third of Japan's Population – and It's Reshaping the Country." *World Economic Forum: Agenda*. September 17, 2019. Accessed on September 22, 2021. https://www.weforum.org/agenda/2019/09/elderly-oldest-population-world-japan/.

Gordon, Deb. "60% Of First-Time Caregivers are Gen Z Or Millennial, New Study Shows." *Forbes*. Feburary 20, 2021. Accessed September 21, 2021. https://www.forbes.com/sites/debgordon/2021/02/20/60-of-first-time-caregivers-are-gen-z-or-millennial-new-study-shows/?sh=757c957e720f.

Kakulla, Brittne. Personal Tech and the Pandemic: Older Adults are Upgrading for a Better Online Experience: 2021 Tech Trends and the 50-Plus: Top 10 Biggest Trends. Washington, DC: AARP Research, September 22, 2021. https://doi.org/10.26419/res.00420.001.

Nahal, Sarbjit and Beijia Ma. "*The Silver Dollar – Longevity Revolution Primer.*" Online: Bank of America Merrill Lynch. June 6, 2014. Accessed September 20, 2021. https://www.longfinance.net/media/documents/The_Silver_Dollar_Longevity_Revolution_Primer.pdf.

OECDiLibrary. "Health at a Glance 2017: OECD Indicators." OECD Publishing. Updated 2017. https://doi.org/10.1787/health_glance-2017-en.

Rothstein, Jake. "Ageing in the Right Place - An Interview With Upsidehom Founder And CEO Jake Rothstein." *The Gerontechnologist*, February 25, 2021. Accessed September 21, 2021. https://www.thegerontechnologist.com/aging-in-the-right-place-an-interview-with-upsidehom-founder-and-ceo-jake-rothstein/.

Schwab, Klaus. "The Fourth Industrial Revolution." *Foreign Affairs*. Updated December 12, 2015. Accessed October 19, 2021. https://

www.foreignaffairs.com/articles/2015-12-12/fourth-industri-al-revolution.

United Nations, Department of Economic and Social Affairs. "World Population Projected to Reach 9.8 Billion in 2050, and 11.2 Billion in 2100." New York, June 21, 2017. https://www.un.org/development/desa/en/news/population/world-popu-lation-prospects-2017.html.

Young, Julie. "Blue Ocean." Investopedia, July 13, 2021. Accessed September 20, 2021. https://www.investopedia.com/terms/b/blue_ocean.asp.

CHAPTER 1: WHAT IS AGING?

Angelou, Maya. *Letter to My Daughter.* New York, New York: Random House Trade Paperbacks, 2008.

ArcGIS. "Media's Role in the Stereotyping of Aging Adults." Accessed October 19, 2021. https://www.arcgis.com/apps/Map-Journal/index.html?appid=136216f5a926499aa7bb7fc3d5a13ca6.

Applewhite, Ashton. *This Chair Rocks: A Manifesto Against Ageism.* New York: Celadon Books, 2020.

Cambridge Dictionary. s.v. "elder," Accessed September 21, 2021. https://dictionary.cambridge.org/dictionary/english/elder.

Costa, Dora L. "Causes of Improving Health and Longevity at Older Ages: A Review of the Explanations." *Genus* 61, no. 1 (2005): 21–38. http://www.jstor.org/stable/29788834.

Emling, Shelley. "The Age at Which You are Officially Old." *AARP.* June 14, 2017. Accessed September 27, 2021. https://www.aarp. org/home-family/friends-family/info-2017/what-age-are-you-old-fd.html.

Ferrucci, Luigi, Marta Gonzalez-Freire, Elisa Fabbri, Eleanor Simonsick, Toshiko Tanaka, Zenobia Moore, Shabnam Salimi, Felipe Sierra, and Rafael de Cabo. "Measuring biological aging in humans: A quest." Aging Cell 19, no. 2, 2019: https://doi. org/10.1111/acel.13080.

Gerbner, George. "Cultivation Analysis: An Overview." Communicator October-December 2000: https://web.asc.upenn.edu/ gerbner/Asset.aspx?assetID=459.

Greenberg, Tamara M. "What Is it Really Like to Be Old?: Can We Truly Be Empathic About Elderly Experience?" *Psychology Today (Blog).* November 15, 2013. Accessed September 19, 2021. https://www.psychologytoday.com/us/blog/21st-century-aging/201311/what-is-it-really-be-old.

Harada, Carolien N., Marissa C. Natelson Love, and Kristen L. Triebel. "Normal Cognitive Aging". *Clinics Geriatric Medicine.* 2013; 29(4):737-752. http://dx.doi.org/10.1016/j.cger.2013.07.002.

Institute for Quality and Efficiency in Health Care. "*What Happens When You Age?*" InformedHealth. Accessed September 23, 2021. https://www.informedhealth.org/what-happens-when-you-age.html.

Jaffe, Ina. "NPR Survey Reveals Despised and Acceptable Terms For Aging." Interviewed by Renee Montagne. Accessed Sep-

tember 24, 2021. https://www.npr.org/2014/07/08/329731428/
npr-poll-reveals-despised-and-acceptable-terms-for-aging.

Kalache, Alexandre and Bruno Lunenfeld. "Men Ageing and
Health Achieving Health Across the Life Span." World Health
Organization. Updated Geneva, April 7, 1999. https://apps.
who.int/iris/bitstream/handle/10665/66941/WHO_NMH_
NPH_01.2.pdf;jsessioni.

Kowal, Paul and Karen Peachey. "*Indicators for the Minimum Data
Set Project on Ageing: A Critical Review in sub-Saharan Africa.*"
WHO. Accessed September 24, 2021. https://www.who.int/
healthinfo/survey/ageing_mds_report_en_daressalaam.pdf.

Lee, Sang Bum, Jae Hun Oh, Jeong Ho Park, Seung Pill Choi, and
Jung Hee Wee. "Differences in Youngest-old, Middle-old, and
Oldest-old Patients Who Visit the Emergency Department."
Clinical and Experimental Emergency Medicine 5, no. 4 (2018):
249–255: https://doi.org/10.15441/ceem.17.261.

Lexico English Dictionary. s.v. "ageing." Accessed September 21,
2021, https://www.lexico.com/definition/Ageing.

Lexico English Dictionary. s.v. "old." Accessed September 21, 2021,
https://www.lexico.com/definition/old.

Löckenhoff, Corinna E., Filip De Fruyt, Antonio Terracciano, et
al. "Perceptions of Aging Across 26 Cultures and Their Cul-
ture-Level Associates." *Psychology and Aging* 24, no. 4 (2009):
941–954. https://doi.org/10.1037/a0016901.

Merriam Webster. s.v. "old." Accessed September 21, 2021, https://www.merriam-webster.com/dictionary/old.

Mevrosh, Sarah. "'Joy, Love, Grief': How It Looks When Families Reunite", *The New York Times,* April 29, 2021. Accessed September 29, 2021. https://www.nytimes.com/2021/04/29/us/coronavirus-nursing-home-reunion.html.

Milken Institute. *"Center for the Future of Aging."* Accessed September 26, 2021. https://milkeninstitute.org/centers/center-for-the-future-of-aging.

MIT AgeLab. "AGNES (Age Gain Now Empathy System)." Accessed September 25, 2021. https://agelab.mit.edu/agnes-age-gain-now-empathy-system.

Nahal, Sarbjit and Beijia Ma. *"The Silver Dollar – Longevity Revolution Primer."* Online: Bank of America Merrill Lynch. Accessed September 20, 2021. https://www.longfinance.net/media/documents/The_Silver_Dollar_Longevity_Revolution_Primer.pdf.

OECD Data. "Elderly Population." Accessed October 5, 2021: https://data.oecd.org/pop/elderly-population.htm.

OECDiLibrary. "Pensions at a Glance 2017: OECD and G20 Indicators." OECD, (Paris, December 5, 2017): 92. http://dx.doi.org/10.1787/pension_glance-2017-en.

Randall, Chris, Abbie Cochrane, Rhian Jones and Silvia Manclossi. "Measuring National Well-Being in the UK: International Comparisons." Office for National Statistics, UK, 2019. https://www.ons.gov.uk/peoplepopulationandcommunity/

wellbeing/articles/measuringnationalwellbeing/internation-
alcomparisons2019.

Pfeffer, Stephanie Emma. "Suzanne Somers, 73, Says She Loves
Aging: 'I've Got a Wisdom That No Young Person Can Buy.'"
People. Accessed September 28, 2021. https://people.com/
health/suzanne-somers-loves-aging-workout-reunion/.

Reinhard, Susan. C, Lynn Friss Feinberg, Ari Houser, Rita Choula,
and Molly Evans. "Valuing the Invaluable: 2019 Update -
Charting a Path Forward." Washington, DC: AARP Public
Policy Institute. November 2019. https://doi.org/10.26419/
ppi.00082.003.

Riberio, Oscar, Laetitia Teixeira, Lia Araujo, and Constança Paúl.
"Caregiver Support Ratio in Europe." *Innovation in Aging* 3, no.
1, November 2019: 138: https://doi.org/10.1093/geroni/igz038.500.

Rogers, Kara, *Encyclopædia Britannica Inc. s.v.* "ageing." Accessed
September 21, 2021, https://www.britannica.com/science/
aging-life-process.

Ruggles, Steven. "Multigenerational Families in Nineteenth-Cen-
tury America." *Continuity and Change* 18, no.1, May, 2003:
139–165. https://doi.org/10.1017/S0268416003004466.

Shira, Dahvi. "Betty White Jokes: I'm 'Much Sexier' at 91 Years
Old Than I Was at 90." *People*. February 5, 2013. Accessed September 26, 2021. https://people.com/tv/betty-white-jokes-im-
much-sexier-at-91-years-old-than-i-was-at-90/.

Taylor, Paul, Morin, Rick, Parker, Kim, Cohn, D'Vera, Wang, Wendy. "Growing Old in America: Expectations vs. Reality." Washington DC, *Pew Research Center*, 2009.

Taylor, Paul, Jeffrey Passe, Richard Fry, Richard Morin, Wendy Wang, Gabriel Velasco, and Daniel Dockterman. "The Return of the Multi-Generational Family Household." *Pew Research Center*, 2010. Accessed October 2, 2021. https://www.pewresearch.org/social-trends/2010/03/18/the-return-of-the-multi-generational-family-household/.

Tedx Talks. "Growing Old: The Unbearable Lightness of Ageing | Jane Caro | TEDxSouthBank." February 9, 2015. Accessed October 2021, Video, 18:47. https://www.youtube.com/watch?v=ULqf3OyemZY.

Thayer, Colette, and Laura Skufca. "Media Image Landscape: Age Representation in Online Images." *AARP Research*, September 2019: https://doi.org/10.26419/res.00339.001.

United Nations, Department of Economic and Social Affairs, Population Division. "World Population Ageing 2015 - Highlights." (ST/ESA/SER.A/368). New York, 2015. https://www.un.org/en/development/desa/population/publications/pdf/ageing/WPA2015_Highlights.pdf.

United Nations, Department of Economic and Social Affairs, Population Division (2017). "World Population Ageing 2017 - Highlights" (ST/ESA/SER.A/397). New York, 2017. Accessed September 23, 2021. https://www.un.org/en/development/desa/population/publications/pdf/ageing/WPA2017_Highlights.pdf.

United Nations Department of Economic and Social Affairs, Population Division (2020). *"World Population Ageing 2020 Highlights: Living Arrangements of Older Persons" (ST/ESA/ SER.A/451).* New York, 2020. https://www.un.org/development/ desa/pd/sites/www.un.org.development.desa.pd/files/undesa_ pd-2020_world_population_ageing_highlights.pdf.

United Nations High Commissioner for Refugees. "Older Persons." UNHCR. June 10, 2021. Accessed September 30, 2021. https:// emergency.unhcr.org/entry/43935/older-persons.

United Nations, "UN Decade of Healthy Ageing 2021 - 2030." WHO. Accessed October 19, 2021. https://www.who.int/ini- tiatives/decade-of-healthy-ageing.

United Nations. "World Fertility Report of 2015." UN.org. New York, 2017. https://www.un.org/development/desa/pd/sites/ www.un.org.development.desa.pd/files/files/documents/2020/ Feb/un_2015_worldfertilityreport_highlights.pdf.

United Nations. "World Population Projected to Reach 9.8 Billion in 2050, and 11.2 Billion in 2100." UN.org. New York , 2017. Accessed September 23, 2021. https://www.un.org/development/ desa/en/news/population/world-population-prospects-2017. html.

Vespa, Jonathan."The US Joins Other Countries With Large Aging Populations." United States Census Bureau. March 13, 2018. Accessed September 22, 2021. https://www.census.gov/library/ stories/2018/03/graying-america.html.

Winfrey, Oprah. *What I Know for Sure.* New York, New York: Flatiron Books, 2014.

CHAPTER 2: WHY DOES THE WORLD NEED AGETECH

AARP Public Policy Institute. "The Aging of the Baby Boom and the Growing Care Gap: A Look at Future Declines in the Availability of Family Caregivers." *Insight on the Issues,* number 85: August 2013. Accessed September 21, 2021. https://www.aarp.org/content/dam/aarp/research/public_policy_institute/ltc/2013/baby-boom-and-the-growing-care-gap-in-brief-AARP-ppi-ltc.pdf.

AOA. "Adult Vision: 41 to 60 Years of Age." American Optometric Association. Accessed October 4, 2021. https://www.aoa.org/healthy-eyes/eye-health-for-life/adult-vision-41-to-60-years-of-age?sso=y.

Anderson, Monica, and Andrew Perrin. "Tech Adoption Climbs Among Older Adults." *Pew Research Center.* Accessed October 5, 2021. https://www.pewresearch.org/internet/2017/05/17/tech-adoption-climbs-among-older-adults/.

Bronswijk, Johanna EMH, Herman Bouma, James L. Fozard, William D. Kearns, Gerald C. Davison, and Pan-Chio Tuan. "Defining Gerontechnology for R&D purposes." *Gerontechnology* 8, no. 1. Winter, 2001: 3–10. https://doi.org/10.4017/gt.2009.08.01.002.00.

Center for Disease Control s.v. "Healthy Places Terminology." Accessed October 2, 2021, https://www.cdc.gov/healthyplaces/terminology.htm.

Cook, Nancy. "Nearly 90% of Americans Age 50 and Older Want to "Age in Place." *Global Newswire*. Accessed October 3, 2021. https://www.globenewswire.com/news-release/2021/05/10/2226492/0/en/NEARLY-90-OF-AMERICANS-AGE-50-AND-OLDER-WANT-TO-AGE-IN-PLACE.html.

CTA Foundation. "OATS CES2021 Interview." January 10, 2021. Video, 18:44. https://www.youtube.com/watch?v=nfx40pRruFw.

Czaja, Sara J., and Chin Chin Lee. "The Impact of Aging on Access to Technology." *Universal Access in the Information Society* 5, no. 4 (December 8, 2007): 341–349. https://doi.org/10.1007/s10209-006-0060-x.

Davey, Judith A., Virginia de Joux, Ganesh Nana, and Mathew Arcus. "Accommodation Options for Older People in Aotearoa/New Zealand." Chrischurch: Centre for Housing Research. June 1, 2004. Accessed October 2, 2021. https://thehub.swa.govt.nz/assets/documents/accommodation_options_for_older_people_in_aotearoa_new_zealand.pdf.

Etkin, Keren. "The Agetech Market Map." TheGerontechnologist.com. Accessed September 29, 2021. https://www.thegerontechnologist.com/wp-content/uploads/2021/03/2021-AgeTech-Market-Map-3.pdf.

Flinn, Brenda. "Millennials: The Emerging Generation of Family Caregivers." AARP Public Policy Institute. Accessed October 3, 2021. https://www.aarp.org/content/dam/aarp/ppi/2018/05/millennial-family-caregivers.pdf.

Fry, Richard. "Millennials overtake Baby Boomers as America's largest generation." *Pew Research Center*. Accessed October 2, 2021. https://www.pewresearch.org/fact-tank/2020/04/28/millennials-overtake-baby-boomers-as-americas-largest-generation/.

Genworth. "Cost of Care Survey." Accessed October 1, 2021. https://www.genworth.com/aging-and-you/finances/cost-of-care.html.

Harada, Carolien N., Marissa C. Natelson Love, Kristen L. Triebel. Normal Cognitive Aging. *Clinics in Geriatric Medicine* 29, no. 4. (2013):737–752. http://dx.doi.org/10.1016/j.cger.2013.07.002.

Indeed Editorial Team. "What Careers Are Most In-Demand Right Now?" Indeed. Accessed October 9, 2021. https://www.indeed.com/career-advice/finding-a-job/in-demand-careers.

Johns Hopkins Medicine. "Age-Related Hearing Loss (Presbycusis)." HEALTH. Accessed October 2, 2021. https://www.hopkinsmedicine.org/health/conditions-and-diseases/presbycusis.

Kakulla, Brittne. *"Personal Tech and the Pandemic: Older Adults are Upgrading for a Better Online Experience: 2021 Tech Trends and the 50-Plus: Top 10 Biggest Trends."* Washington, DC: AARP Research, September 22, 2021. https://doi.org/10.26419/res.00420.001.

Kamber, Thomas, Alex Glazebrook, Clarence Burton, Aaron Osgood, Suzanne Myklebust, Melissa Sakow, DeAnne Cuellar, Rita Soni, and Joyce Weil. "Aging Connected: Closing the Connectivity Gap for Older Americans." Older Adults Technology Services and the Humana Foundation (2021). Accessed October 5, 2021, https://agingconnected.org/wp-content/uploads/2021/01/Aging-Connected_2021.pdf.

Köttl, Hanna, and Ittay Mannheim. "Ageism & Digital Technology: Policy Measures to Address Ageism as a Barrier to Adoption and Use of Digital Technology." Euro Ageism. Accessed October 5, 2021. https://euroageism.eu/wp-content/uploads/2021/03/Ageism-and-Technology-Policy-Brief.pdf.

LaBerge, Laura, Clayton O'Toole, Jeremy Schneider, and Kate Smaje. "How COVID-19 Has Pushed Companies Over the Technology Tipping Point—And Transformed Business Forever." Mckinsey & Company. Edited by Daniella Seilar. Accessed October 4, 2021. https://www.mckinsey.com/business-functions/strategy-and-corporate-finance/our-insights/how-covid-19-has-pushed-companies-over-the-technology-tipping-point-and-transformed-business-forever.

Merriam Webster. s.v. "technology." Accessed October 5, 2021. https://www.merriam-webster.com/dictionary/technology.

Nahal, Sarbjit and Beijia Ma. "The Silver Dollar – longevity revolution primer." Online: Bank of America Merrill Lynch. June 6, 2014. Accessed September 20, 2021. https://www.longfinance.net/media/documents/The_Silver_Dollar_Longevity_Revolution_Primer.pdf.

Norman, Don and Bruce Tognazzini. "How Apple Is Giving Design a Bad Name: For years, Apple Followed User-centered Design Principles. Then Something Went Wrong." *Fast Company.* November 10, 2015. Accessed October 6, 2021. https://

www.fastcompany.com/3053406/how-apple-is-giving-design-a-bad-name.

Norman, Don. "I Wrote the Book on User-Friendly Design. What I See Today Horrifies Me: The World Is Designed Against the Elderly, Writes Don Norman, 83-Year-Old Author of the Industry Bible Design of Everyday Things and A Former Apple VP." *Fast Company.* August 5, 2019. Accessed October 4, 2021. https://www.fastcompany.com/90338379/i-wrote-the-book-on-user-friendly-design-what-i-see-today-horrifies-me.

OECDiLibrary. "Understanding the Digital Divide." OECD Publications. August 5, 2002. https://stats.oecd.org/glossary/detail.asp?ID=4719.

Osterman, Paul. "Who Will Care for Us? Long-Term Care and the Long-Term Workforce." Russell Sage Foundation, New York 2017. https://doi.org/10.7758/9781610448673.

Pew Research Center. "Internet/Broadband Fact Sheet." Accessed October 4, 2021. https://www.pewresearch.org/internet/fact-sheet/internet-broadband/.

"About SeniorNet." SeniorNet. Accessed October 19, 2021. https://seniornet.org/index.php/about-us/.

Vogels, Emily A. "Millennials stand out for their technology use, but older generations also embrace digital life." *Pew Research Center.* Accessed October 2, 2021. https://www.pewresearch.org/fact-tank/2019/09/09/us-generations-technology-use/.

World Wide Web Foundation. "History of the Web." Accessed October 2, 2021. https://webfoundation.org/about/vision/history-of-the-web/.

Wu, Ya-Huei, Manon Lewis, and Anne-Sophie Rigaud. "Cognitive Function and Digital Device Use in Older Adults Attending a Memory Clinic." *Gerontology and Geriatric Medicine* 5 (May 2, 2019): 1–7. https://doi.org/10.1177/2333721419844886.

Zip Recruiter. "What Is the Average Caregiver Salary by State." Accessed October 7, 2021. https://www.ziprecruiter.com/Salaries/What-Is-the-Average-Caregiver-Salary-by-State.

CHAPTER 3: WHY NOW?

Accius, Jean, and Joo Yeoun Suh. "The Longevity Economy Outlook: How People Ages 50 and Older Are Fueling Economic Growth, Stimulating Jobs, and Creating Opportunities for All." Washington, DC: AARP Thought Leadership, December 2019. https://doi.org/10.26419/int.00042.001.

Adamczyk, Alicia. "Millennials Own Less Than 5% of All U.S. Wealth." CNBC. October 9, 2020. Accessed October 1, 2021. https://www.cnbc.com/2020/10/09/millennials-own-less-than-5percent-of-all-us-wealth.html.

Anderson, Janna, and Lee Rainie. "The Negatives of Digital Life." Pew Research Center, July 3, 2018. Accessed September 30, 2021. https://www.pewresearch.org/internet/2018/07/03/the-negatives-of-digital-life/.

Anderson, Janna, and Lee Rainie. "The Positives of Digital Life." Pew Research Center. Accessed October 1, 2021. https://www.pewresearch.org/internet/2018/07/03/the-positives-of-digital-life/.

Bono, Sal. "How Technology Is Isolating Our Elderly and How to Fix It." *Inside Edition*. April 29, 2021. Accessed September 29, 2021. https://www.insideedition.com/how-technology-is-isolating-our-elderly-and-how-to-fix-it-66490.

Deloitte Touche Tohmatsu Limited. "The Digital Transformation of Customer Services." Deloitte. Accessed September 28, 2021. https://www2.deloitte.com/content/dam/Deloitte/nl/Documents/technology/deloitte-nl-paper-digital-transformation-of-customer-services.pdf.

Galloway, Scott. "This Is the Best Time to Start a Business: The Economic Flood Gates are About to Open Wide." *Medium. No Mercy, No Malice (blog)*, March 30, 2021. Accessed October 1, 2021. https://marker.medium.com/scott-galloway-this-is-the-best-time-to-start-a-business-2ca386ac73e0.

Kakulla, Brittne. "Personal Tech and the Pandemic: Older Adults are Upgrading for a Better Online Experience: 2021 Tech Trends and the 50-Plus: Top 10 Biggest Trends." Washington, DC: AARP Research, September 22, 2021. https://doi.org/10.26419/res.00420.001.

Kamber, Thomas, Alex Glazebrook, Clarence Burton, Aaron Osgood, Suzanne Myklebust, Melissa Sakow, DeAnne Cuellar, Rita Soni, and Joyce Weil. "Aging Connected: Closing the Connectivity Gap for Older Americans." Older Adults Technology Services and the Humana Foundation (2021), Accessed October 5, 2021, https://agingconnected.org/wp-content/uploads/2021/01/Aging-Connected_2021.pdf.

Kridel, Tim. "Senior Living Communities Upgrade Networks to Support New Devices." *Health Tech Magazine,* November 4, 2019. https://healthtechmagazine.net/article/2019/11/senior-living-communities-upgrade-networks-support-new-devices.

LaBerge, Laura, Clayton O'Toole, Jeremy Schneider, and Kate Smaje. "How COVID-19 Has Pushed Companies Over the Technology Tipping Point—And Transformed Business Forever." Mckinsey & Company, October 5, 2020. Edited by Daniella Seilar. Accessed October 4, 2021. https://www.mckinsey.com/business-functions/strategy-and-corporate-finance/our-insights/how-covid-19-has-pushed-companies-over-the-technology-tipping-point-and-transformed-business-forever.

Levy, Steven, *Britannica, s.v.* "Apple Inc." Accessed October 2021. https://www.britannica.com/topic/Apple-Inc.

Nahal, Sarbjit and Beijia Ma. "The Silver Dollar – Longevity Revolution Primer." Online: Bank of America Merrill Lynch. Accessed September 20, 2021. https://www.longfinance.net/media/documents/The_Silver_Dollar_Longevity_Revolution_Primer.pdf.

Salesforce. "What Is Digital Transformation?" Accessed September 29, 2021. https://www.salesforce.com/eu/products/platform/what-is-digital-transformation/.

Schwab, Klaus. "The Fourth Industrial Revolution: What It Means, How to Respond." World Economic Forum. Accessed September 29, 2021. https://www.weforum.org/agenda/2016/01/the-fourth-industrial-revolution-what-it-means-and-how-to-respond/.

Stone, Will. "'Just Cruel': Digital Race For COVID-19 Vaccines Leaves Many Seniors Behind." *NPR*. Accessed October 3, 2021. https://www.npr.org/sections/health-shots/2021/02/04/963758458/digital-race-for-covid-19-vaccines-leaves-many-seniors-behind.

United Nations, "UN Decade of Healthy Ageing 2021 - 2030. WHO, Accessed October 19,2021. https://www.who.int/initiatives/decade-of-healthy-ageing.

CHAPTER 4: THE CHALLENGES OF AGING

Bloom, David E., Axel Boersch-Supan, Patrick McGee, and Atsushi Seike. "Population Aging: Facts, Challenges, and Responses." *Benefits and Compensation International* 41, no. 1 (2011): 22.

Centers for Disease Control Prevention. "Keep on Your Feet—Preventing Older Adult Falls." December 16, 2020. Accessed October 12, 2021. https://www.cdc.gov/injury/features/older-adult-falls/.

Collinson, Catherine, Patti Rowey, and Heidi Cho. "Retirement Security Amid COVID-19: The Outlook of Three Generations." Transamerica for Retirement Studies. Accessed October 7, 2021. https://transamericacenter.org/docs/default-source/

retirement-survey-of-workers/tcrs2020_sr_retirement_security_amid_covid-19.pdf.

Crunchbase. "Seed Round - Bold." BOLD. Accessed October 11, 2021. https://www.crunchbase.com/organization/bold-b06e.

Dersarkissian, Carol. "Sarcopenia With Aging." *WebMD*. Accessed October 11, 2021. https://www.webmd.com/healthy-aging/guide/sarcopenia-with-aging.

Edemekong, Peter F., Deb L. Bomgaars, Sukesh Sukumaran, and Shoshana B. Levy. "Activities of Daily Living." StatPearls Publishing, Treasure Island (FL), January 2021. https://www.ncbi.nlm.nih.gov/books/NBK470404/.

Enda, Grace and William G. Gale. "How Does Gender Equality Affect Women in Retirement?" Brookings. Accessed October 6, 2021. https://www.brookings.edu/essay/how-does-gender-equality-affect-women-in-retirement/.

Eshkoor, Sima Ataollahi, Tengku Aizan Hamid, Chan Yoke Mun, and Chee Kyun Ng. "Mild Cognitive Impairment and Its Management in Older People." *Clinical Interventions in Aging* 10 (April 10, 2015): 687. https://doi.org/10.2147/CIA.S73922.

Fahle, Sean and Kathleen McGarry. "5. Women Working Longer: Labor Market Implications of Providing Family Care." In *Women Working Longer: Increased Employment at Older Ages*. Edited by Claudia Goldin and Lawrence F. Katz, 157–182. Chicago: University of Chicago Press, 2018. https://doi.org/10.7208/9780226532646-007.

Fidelity. "How to Plan for Rising Health Care Costs." Accessed October 5, 2021. https://www.fidelity.com/viewpoints/personal-finance/plan-for-rising-health-care-costs.

Fox, Michelle. "Nearly 1 in 5 Working Women Have Nothing Saved for Retirement." *CNBC*. Accessed October 9, 2021. https://www.cnbc.com/2020/03/05/nearly-1-in-5-working-women-have-nothing-saved-for-retirement.html.

Genworth. "Cost of Care Survey, 2020." Accessed October 4, 2021. https://www.genworth.com/aging-and-you/finances/cost-of-care.html.

Gosselin, Peter. "Age Discrimination: If You're Over 50, Chances Are the Decision to Leave a Job Won't be Yours." *ProPublica*. Accessed October 7, 2021. https://www.propublica.org/article/older-workers-united-states-pushed-out-of-work-forced-retirement.

Hood, Bryan. "The World's First Robot Chef Is Finally Here, and It Even Cleans Up After Itself: The Moley Robotic Kitchen will Come Pre-programmed with over 5,000 Recipes." Robbreport. Accessed October 9, 2021. https://robbreport.com/gear/electronics/moley-robotics-robot-kitchen-uk-for-sale-1234590791/.

Langa KM. "Cognitive Aging, Dementia, and the Future of an Aging Population." National Center for Biotechnology. Edited by Majmundar MK, Hayward MD. Washington: National Academies Press, 2018. https://www.ncbi.nlm.nih.gov/books/NBK513075/.

Lee, Dami. "This $16,000 robot uses artificial intelligence to sort and fold laundry." *The Verge.* Accessed October 9, 2021. https://www.theverge.com/2018/1/10/16865506/laundroid-laundry-folding-machine-foldimate-ces-2018.

Matsui, Kathy, Hiromi Suzuki, and Kazunori Tatebe. "Womenomics 5.0." The Goldman Sachs Group, Inc. Accessed October 4, 2021. https://www.goldmansachs.com/insights/pages/womenomics-5.0/multimedia/womenomics-5.0-report.pdf.

McKhann, GM, DS Knopman, H Chertkow, et al. "The Diagnosis of Dementia Due to Alzheimer's Disease: Recommendations from the National Institute on Aging–Alzheimer's Association Workgroups on Diagnostic Guidelines for Alzheimer's Disease.' *Alzheimer's & Dementia* 7, no. 3 (April 22, 2011): 263–269. https://doi.org/10.1016/j.jalz.2011.03.005.

Osterman, Paul. *Who Will Care for Us? Long-Term Care and the Long-Term Workforce.* Russell Sage Foundation. New York, 2017. https://doi.org/10.7758/9781610448673.

Pomeroy, Claire. "Loneliness Is Harmful to Our Nation's Health: Research Underscores the Role of Social Isolation in Disease and Mortality." Scientific American, March 20, 2019. Accessed October 12, 2021. https://blogs.scientificamerican.com/observations/loneliness-is-harmful-to-our-nations-health/.

Riberio, Oscar, Laetitia Teixeira, Lia Araujo, and Constança Paúl. "Caregiver Support Ratio in Europe." *Innovation in Aging* 3, no. 1, November 2019: 138. https://doi.org/10.1093/geroni/igz038.500.

Sawyer, Bradley and Gary Claxton. "How Do Health Expenditures Vary Across the Population?" Peterson-KFF Health Systems Tracker. Accessed October 11, 2021. https://www.healthsystemtracker.org/chart-collection/health-expenditures-vary-across-population/#item-start.

Stasha, Smiljanic. "The State of Healthcare Industry – Statistics for 2021." *Policy Advice*. August 6, 2021. Accessed October 11, 2021. https://policyadvice.net/insurance/insights/healthcare-statistics/.

Task Force on Research and Development for Technology to Support Aging Adults Committee on Technology of the National Science & Technology Council. "Emerging Technologies to Support an Aging Population." Trump White House Archives. March, 2019. https://trumpwhitehouse.archives.gov/wp-content/uploads/2019/03/Emerging-Tech-to-Support-Aging-2019.pdf.

The Global Wellness Institute. "What is Wellness?" Accessed October 10, 2021. https://globalwellnessinstitute.org/what-is-wellness.

United Nations, Department of Economic and Social Affairs, Population Division. "World Population Ageing 2017 - Highlights." (ST/ESA/SER.A/397), New York, 2017. Accessed September 23, 2021. https://www.un.org/en/development/desa/population/publications/pdf/ageing/WPA2017_Highlights.pdf.

US Department of Health and Human Services, Centers for Disease Control and Prevention, National Center for Health Statistics. "Early Release of Selected Estimates Based on Data from the National Health Interview Survey, 2016." CDC, 2016. Accessed October 7, 2021. https://www.cdc.gov/nchs/data/nhis/earlyrelease/earlyrelease201705_12.pdf.

Wilkinson, Jens. "The Strong Robot with the Gentle Touch." Riken. Accessed October 5, 2021. https://www.riken.jp/en/news_pubs/research_news/pr/2015/20150223_2/.

World Health Organization, "Dementia." September 2, 2021. Accessed October4, 2021. https://www.who.int/news-room/fact-sheets/detail/dementia.

CHAPTER 5: A CLOSER LOOK AT MOBILITY AND TRANSPORTATION

Aging Gracefully across Environments to Ensure Well-Being, Engagement and Long life. "How ROSA the Service Robot Will Help Isolated Seniors and Support Aging in Place." AGE-WELL news. Accessed September 22, 2021. https://agewell-nce.ca/archives/10207.

Kakulla, Brittne. "Personal Tech and the Pandemic: Older Adults Are Upgrading for a Better Online Experience: 2021 Tech Trends and the 50-Plus: Top 10 Biggest Trends." Washington, DC: AARP Research, September 22, 2021. https://doi.org/10.26419/res.00420.001.

Korosec, Kristen. "Tesla Refutes Elon Musk's Timeline on 'Full Self-Driving.'" *TechCrunch,* May 8, 2021. Accessed October 14, 2021. https://techcrunch.com/2021/05/07/tesla-refutes-elon-musks-timeline-on-full-self-driving/.

Sudo, Chuck. "Senior Communities Become Key Testing Ground for Self-Driving Cars." *Senior Housing News,* March 17, 2019. Accessed October 12, 2021. https://seniorhousingnews. com/2019/03/17/senior-communities-become-key-testing-ground-for-self-driving-cars/.

CHAPTER 6: THERE'S NO PLACE LIKE HOME

Bowers, Lois. "Senior Living Occupancy Reaches New Record Low of 78.8 Percent in First Quarter: NIC." *Mcknight's Senior Living.* Accessed October 2021. https://www.mcknightsseniorliving. com/home/news/senior-living-occupancy-reaches-new-record-low-of-78-8-percent-in-first-quarter-nic/.

Breeding, Brad. "The Evolution of Senior Living in the U.S." *My Life Site.* Accessed October 15, 2021. https://mylifesite.net/blog/post/the-evolution-of-senior-living-in-the-u-s/.

Comas-Herrera, Adelina, Joseba Zalakaín, Elizabeth Lemmon, David Henderson, Charles Litwin, Amy T. Hsu, Andrea E. Schmidt, Greg Arling Florien Kruse, and Jose-Luis Fernández. "Mortality Associated with COVID-19 in Care Homes: International Evidence." LTCcovid.org. April 12, 2020.

Ecker, Elizabeth. "Three Trends to Watch in Senior Living Tech This Year." *Senior Housing News.* Accessed October 17, 2021. https://seniorhousingnews.com/2021/04/06/three-trends-to-watch-in-senior-living-tech-this-year/.

Hoyt, Jeff. "History of Senior Living." Seniorliving.org. Accessed October 12, 2021. https://www.seniorliving.org/history/.

Irving, Paul. "The Villages Is a Success Story, But Many of Us Want Something Different in a Place to Live as We Get Older." *Next Avenue*. Accessed October 2021. https://www.nextavenue.org/the-villages-lacks-diversity/.

Jacobs, Karrie. "Don't Mind the Gap in Intergenerational Housing." *The New York Times*. Accessed October 2021. https://www.nytimes.com/2021/09/02/style/housing-elderly-intergenerational-living.html.

Mullaney, Tim. "With New CEO, Formation Capital Launches Strategy to 'Break Down Walls of Senior Living'." *Senior Housing News*. Accessed October 11, 2021. https://seniorhousingnews.com/2021/05/01/with-new-ceo-formation-capital-launches-strategy-to-break-down-walls-of-senior-living/.

NIC. "NIC Talks 2019 | Jody Holtzman | Longevity Venture Advisors." September 18, 2019. Video, 13:00. https://www.youtube.com/watch?v=v-2ZLS-aQkU&t=18s.

PBS NewsHour. "How More Americans Are 'Aging in Place.'" November 6, 2013. Video, 11:52. https://www.youtube.com/watch?v=5VqgSkN14JE.

Regan, Tim. "COVID-19 Opened the Telehealth Floodgates, But Barriers Remain in Senior Living." *Senior Housing News*. Accessed October 16, 2021. https://seniorhousingnews.com/2020/06/24/covid-19-opened-the-telehealth-floodgates-but-barriers-remain-in-senior-living/.

Taylor, Paul, Jeffrey Passe, Richard Fry, Richard Morin, Wendy Wang, Gabriel Velasco, and Daniel Dockterman. "The Return of the Multi-Generational Family Household." *Pew Research Center*, 2010. Accessed October 2, 2021. https://www.pewresearch.org/social-trends/2010/03/18/the-return-of-the-multi-generational-family-household/.

Tyronda, James. "'Granny Flats' Could Be Answer for Many Family Caregivers." *WXXI News*. Accessed October 11, 2021. https://www.wxxinews.org/post/granny-flats-could-be-answer-many-family-caregivers.

Van Leeuwen, K. M., G. Widdershoven, R. Ostelo, et al. "What Does Quality of Life Mean to Older Adults? A Thematic Synthesis." *PloS one* 14, no. 3 (March, 2019). https://doi.org/10.1371/journal.pone.0213263.

CHAPTER 7: ABOUT BUILDING DIGITAL PRODUCTS FOR OLDER ADULTS

Bennett, John, Laura De Young, and Bradley Hartfield. "*Bringing Design to Software.*" Edited by Terry Winograd. Association for Computing Machinery. New York, 1996. https://doi.org/10.1145/229868.

Collins Dictionary. s.v. "JND." Accessed September 23, 2021. https://www.collinsdictionary.com/dictionary/english/jnd.

Coughlin, Joseph F. and Luke Yoquinto. "Technology for Older People Doesn't Have to Be Ugly." *Wall Street Journal,* October 14, 2018. https://www.wsj.com/articles/technology-for-older-people-doesnt-have-to-be-ugly-1539546423.

Famakinwa, Joyce. "Papa Steps into the Virtual Care Arena with Launch of 'Papa Health." *Home Healthcare News*. Accessed October 11, 2021. https://homehealthcarenews.com/2020/12/papa-steps-into-the-virtual-care-arena-with-launch-of-papa-health/.

Farr, Christina. "Private Medicare Plan Devoted Health Says It Is the First to Cover Apple Watch as a Benefit." *CNBC*. Accessed October 10, 2021. https://www.cnbc.com/2019/10/07/devoted-medicare-advantage-plan-covering-apple-watch-as-a-benefit.html.

Golant, Stephen M. "A Theoretical Model to Explain the Smart Technology Adoption Behaviors of Elder Consumers (Elderadopt)." *Journal of Aging Studies* 42 (August, 2017): 56-73. https://doi.org/10.1016/j.jaging.2017.07.003.

Gomes, Lee. "Apple's Scientists Leave as Cuts In R&D Take Toll on Research." *Wall Street Journal*, 1997. https://www.wsj.com/articles/SB859765666689891000.

Mitzner, Tracy L., Jyoti Savla, Walter R. Boot, Joseph Sharit, Neil Charness, Sara J. Czaja, and Wendy A Rogers. "Technology Adoption by Older Adults: Findings From the PRISM Trial." *The Gerontologist*, Volume 59, Issue 1, (February 2019): 34–44. https://doi.org/10.1093/geront/gny113.

Norman, Donald. "Design as Practiced." Stanford University. Accessed October 11, 2021. https://hci.stanford.edu/publications/bds/12-norman.html.

Norman, Don and Bruce Tognazzini. "How Apple Is Giving Design a Bad Name: For years, Apple followed user-centered design principles. Then something went wrong." Fast Company. Accessed October 6, 2021. https://www.fastcompany.com/3053406/how-apple-is-giving-design-a-bad-name.

Norman, Don. "I Wrote the Book on User-Friendly Design. What I See Today Horrifies Me: The World Is Designed Against the Elderly, Writes Don Norman, 83-Year-Old Author of the Industry Bible Design of Everyday Things and A Former Apple VP." Fast Company. Accessed October 4, 2021. https://www.fastcompany.com/90338379/i-wrote-the-book-on-user-friendly-design-what-i-see-today-horrifies-me.

Papa Health. Home Page. Accessed October 17, 2021. https://www.papa.health/.

Rose, Don. "Life Alert's Slogan 'I've Fallen, And I Can't Get Up!®' Ranked Number One on USA TODAY's List of Most Memorable Ad Campaigns." PRWeb. Accessed October 12, 2021. https://www.prweb.com/releases/lifealert/usatoday/prweb544512.htm.

Taherdoost, Hamed. "A Review of Technology Acceptance and Adoption Models and Theories." Procedia Manufacturing 22 (2018): 960–967. https://doi.org/10.1016/j.promfg.2018.03.137.

World Health Organization. "Social Determinants of Health." Accessed September 23, 2021, https://www.who.int/health-topics/social-determinants-of-health#tab=tab_1.

CHAPTER 8: ABOUT THE ROLE OF GOVERNMENTS AND NGOS

Arcelus, Dr. Amaya. "Aging in Place Challenge Program." National Research Council Canada. Accessed October 4, 2021. https://nrc.canada.ca/en/research-development/research-collaboration/programs/aging-place-challenge-program.

Carman, Ashley. "Jibo, The Social Robot That Was Supposed to Die, Is Getting a Second Life." *The Verge*. Accessed October 15, 2021. https://www.theverge.com/2020/7/23/21325644/jibo-social-robot-ntt-disruptionfunding.

Eurostat. "Population Structure and Ageing." Accessed October 2021. https://ec.europa.eu/eurostat/statistics-explained/index.php?title=Population_structure_and_ageing#The_share_of_elderly_people_continues_to_increase.

Federal Communications Commission. "Emergency Broadband Benefit." FCC. Accessed October 11, 2021. https://www.fcc.gov/broadbandbenefit.

Holly, Robert. "National Home Health Spending Reaches All-Time High of $113.5 Billion." *Home Healthcare News*. Accessed October 4, 2021. https://homehealthcarenews.com/2020/12/national-home-health-spending-reaches-all-time-high-of-113-5-billion/.

Holly, Robert. "Caregiver Turnover Rate Falls to 64% as Home Care Agencies 'Flatten the Curve'." *Home Healthcare News*. Accessed October 2, 2021. https://homehealthcarenews.com/2020/06/caregiver-turnover-rate-falls-to-64-as-home-care-agencies-flatten-the-curve/.

Horizon 2020. "Health, Demographic Change and Wellbeing." Accessed October 9, 2021. https://ec.europa.eu/programmes/ horizon2020/en/h2020-section/health-demographic-change-and-wellbeing.

IBISworld. "Home Care Providers in the US - Market Size 2002–2027." Accessed October 5, 2021. https://www.ibisworld.com/ industry-statistics/market-size/home-care-providers-united-states/.

Israel Innovation Authority. Home Page. Accessed October 17, 2021. https://innovationisrael.org.il/en/.

Israel Innovation Authority. "Entrepreneur." Innovation Israel. Accessed October 16, 2021. https://innovationisrael.org.il/en/ our-value-propositions/entrepreneur.

Jacobson, Julie. "Big Win for Home Health Technology: Medicare, Medicaid to Up Reimbursement." *CEPro*. Accessed September 27, 2021. https://www.cepro.com/news/connected_health_centers_for_medicare_medicaid_reimbursement_health_tech/.

Kakulla, Brittne. "Personal Tech and the Pandemic: Older Adults Are Upgrading for a Better Online Experience: 2021 Tech Trends and the 50-Plus: Top 10 Biggest Trends." Washington, DC: AARP Research, September 22, 2021. https://doi. org/10.26419/res.00420.001.

Kamber, Thomas, Alex Glazebrook, Clarence Burton, Aaron Osgood, Suzanne Myklebust, Melissa Sakow, DeAnne Cuellar, Rita Soni, and Joyce Weil. "Aging Connected: Closing the Connectivity Gap for Older Americans, Older Adults Technology

Services and the Humana Foundation." Accessed October 5, 2021. https://agingconnected.org/wp-content/uploads/2021/01/Aging-Connected_2021.pdf.

Kempton, Misty. "How Much Is Caregiver Turnover Really Costing Your Business?" *Home Care Pulse.* Accessed October 8, 2021. https://www.homecarepulse.com/articles/much-caregiver-turnover-really-costing-business/.

Loughran, Jack. "UK Government Endorses Elderly Tech to Give People Five More Years of Independent Living." Engineering and Technology. Accessed October 5, 2021. https://eandt.theiet.org/content/articles/2019/07/uk-government-endorses-elderly-tech-to-give-people-five-more-years-of-independent-living/.

Medicaid. "Money Follows the Person." Medicaid.gov. Accessed September 25, 2021. https://www.medicaid.gov/medicaid/long-term-services-supports/money-follows-person/index.html.

Royal Commission into Aged Care Quality and Safety. "Final Report of the Royal Commission into Aged Care Quality and Safety." Accessed October 5, 2021. https://agedcare.royalcommission.gov.au/publications/final-report.

Silver, Caleb. "The Top 25 Economies in the World." *Investopedia,* December 24, 2020. Accessed October 6, 2021. https://www.investopedia.com/insights/worlds-top-economies/.

The White House. "Fact Sheet: The American Jobs Plan." Accessed October 11, 2021. https://www.whitehouse.gov/briefing-room/statements-releases/2021/03/31/fact-sheet-the-american-jobs-plan/.

Trudeau, Justin. "Connecting all Canadians to high-speed Internet." Prime Minister of Canada. November 9, 2020. Accessed October 5, 2021. https://pm.gc.ca/en/news/news-releases/2020/11/09/connecting-all-canadians-high-speed-internet.

United Nations. "UN Decade of Healthy Ageing 2021 - 2030." WHO. Accessed October 9, 2021. https://www.who.int/initiatives/decade-of-healthy-ageing.

Yeginsu, Ceylan. "UK Appoints a Minister for Loneliness." *The New York Times.* Accessed October 2, 2021. https://www.nytimes.com/2018/01/17/world/europe/uk-britain-loneliness.html.

תוכנית "מאיצי עדם" לקידוס מיזמים מזהי דודיעו תומז בקרב סירקוח היימדקאב ובכמונ, היגולונכטהו עדמה תונשדחו, דרשמ רקחמה, September 5, 2021. Accessed October 17, 2021. https://www.gov.il/he/departments/publications/Call_for_bids/rfp20210905.

CHAPTER 9: FUTURE OPPORTUNITIES

AARP. "Tech Usage Among Older Adults Skyrockets During Pandemic: New AARP Research Report Highlights Rise in Tech Ownership, Adoption, and Usage, Yet Access to Internet is Still an Issue for Older Americans" Press AARP. Accessed October 1, 2021. https://press.aarp.org/2021-4-21-Tech-Usage-Among-Older-Adults-Skyrockets-During-Pandemic.

Age Diversity. "Older Workers' Outdated Skills and Resistance to Retraining." The Center for Research into the Older Workforce. Accessed September 20, 2021. https://www.agediversity.org/

course/older-workers-outdated-skills-and-resistance-to-re-training/.

Akinola, Sofiat. "How Can We Best Engage Older Workers in Reskilling Efforts?" *World Economic Forum: Agenda*. Accessed October 10, 2021. https://www.weforum.org/agenda/2021/05/how-can-we-engage-older-workers-in-reskilling-efforts-jobs-reset-summit-ageing-workforce-longevity-upskilling/.

Bestsennyy, Oleg, Greg Gilbert, Alex Harris, and Jennifer Rost. "Telehealth: A Quarter-Trillion-Dollar Post-COVID-19 Reality?" Mckinsey & Company. Accessed October 10, 2021. https://www.mckinsey.com/industries/healthcare-systems-and-services/our-insights/telehealth-a-quarter-trillion-dollar-post-covid-19-reality.

Carrington, Damian. "No-kill, Lab-Grown Meat to Go on Sale for First Time." *The Guardian*. Accessed October 10, 2021. https://www.theguardian.com/environment/2020/dec/02/no-kill-lab-grown-meat-to-go-on-sale-for-first-time.

Cominetti, Nye. "A U-shaped crisis: The Impact of the COVID-19 Crisis on Older Workers." Resolution Foundation. Accessed October 7, 2021. https://www.resolutionfoundation.org/app/uploads/2021/04/A-U-shaped-crisis.pdf.

De Luce, Ivan. "SkyDrive and Uber Air could have flying cars in the skies by 2023." The Business of Business. Accessed October 9, 2021. https://www.businessofbusiness.com/articles/skydrive-boeing-airbus-uber-air-flying-cars-data/.

Dezern, Deanna. Interview by Natalie Hoke. "What's it Like to Live with ElliQ? Video Interview with User Deanna." *ElliQ*. Accessed October 8, 2021. https://blog.elliq.com/whats-it-like-to-live-with-elliq-video-interview-with-user-deanna.

Dychtwald, Ken, Tamara J. Erickson, and Bob Morison. "It's Time to Retire Retirement." *Harvard Business Review*. Accessed October 8, 2021. https://hbr.org/2004/03/its-time-to-retire-retirement.

Edmond, Charlotte. "Elderly People Make Up a Third of Japan's Population – and It's Reshaping the Country." World Economic Forum: Agenda. Accessed on September 22, 2021. https://www.weforum.org/agenda/2019/09/elderly-oldest-population-world-japan/.

Galloway, Scott. "This Is the Best Time to Start a Business: The Economic Flood Gates are About to Open Wide." *Medium. No Mercy, No Malice*. Accessed October 1, 2021. https://marker.medium.com/scott-galloway-this-is-the-best-time-to-start-a-business-2ca386ac73e0.

Greeley, Greg. "Ageless Travel: The Growing Popularity of Airbnb for the Over 60s." Airbnb. Accessed October 5, 2021. https://news.airbnb.com/ageless-travel-the-growing-popularity-of-airbnb-for-the-over-60s/.

Grossman, David. "How Do NASA's Apollo Computers Stack Up to an iPhone?: Better Than You Think." *Popular Mechanics*. Accessed October 8, 2021. https://www.popularmechanics.com/space/moon-mars/a25655/nasa-computer-iphone-comparison/.

Marcus, Jon. "In One Year, Pandemic Forced Millions of Workers to Retire Early: From Flight Attendants to Grocery Store Managers, Older Adults Made the Tough Decision to End Careers." *AARP: Working 50+*. Accessed October 10, 2021. https://www. aarp.org/work/working-at-50-plus/info-2021/pandemic-workers-early-retirement.html.

McLeod, Saul. "Maslow's Hierarchy of Needs." *Simply Psychology* 1, no. 1–18 (2007). Accessed October 8, 2021. https://www.simplypsychology.org/maslow.html.

National Academies of Sciences, Engineering, and Medicine. "Social Isolation and Loneliness in Older Adults: Opportunities for the Health Care System." The National Academies Press, February 27, 2020. https://doi.org/10.17226/25663.

Rouzet, Dorothée, Aida Caldera Sánchez, Théodore Renault, and Oliver Roehn. "Fiscal Challenges and Inclusive Growth in Ageing Societies." Authorized for publication by Laurence Boone Chief Economist and G20 Finance Deputy. *OECD Economic Policy Papers*, No. 27, OECD Publishing, (Paris, September 10, 2019), https://doi.org/10.1787/c553d8d2-en.

Santacreu, Ana Maria. "Long-Run Economic Effects of Changes in the Age Dependency Ratio." *Economic Synopses*, No. 17, 2016. https://doi.org/10.20955/es.2016.17.

Ship-Technology. "How Old Is the Average Cruise Passenger?" Accessed October 8, 2021. https://www.ship-technology.com/features/how-old-is-the-average-cruise-passenger/.

Solis-Moreira, Jocelyn. "How Did We Develop a COVID-19 Vaccine so Quickly? *Medical News Today.* Accessed October 7, 2021. https://www.medicalnewstoday.com/articles/how-did-we-develop-a-covid-19-vaccine-so-quickly.

Suciu, Peter. "Seniors Are Using Dating Apps and Tinder Leads The Pack." *Forbes.* Accessed October 6, 2021. https://www.forbes.com/sites/petersuciu/2021/05/14/seniors-are-using-dating-apps-and-tinder-leads-the-pack/?sh=1e2288da2dd5.

Tillman, Maggie. "What is Google Duplex, Where Is It Available, and How Does It Work?" Pocket-lint. Accessed September 29, 2021. https://www.pocket-lint.com/phones/news/google/146008-what-is-google-duplex-where-is-it-available-and-how-does-it-work.

Weller, Christian. "Even Amid Low Unemployment, Many Workers Struggle to Find a Job." *Forbes.* Accessed October 2, 2021. https://www.forbes.com/sites/christianweller/2019/10/08/even-amid-low-unemployment-many-workers-still-struggle-finding-a-job/?sh=180b38a94de2.

CHAPTER 10: A POSSIBLE FUTURE

Insurance Institute for Highway Safety. "Fatality Facts 2019 Pedestrians." *IIHS.* Accessed October 5, 2021. https://www.iihs.org/topics/fatality-statistics/detail/pedestrians.